FORGED IN FIRE

David Ferrier

Battle Press
SATELLITE BEACH, FLORIDA

Battle Press books may be ordered through booksellers or
by contacting:

Battle Press
1-919-218-4039
steve@battlepress.media
www.battlepress.media

ISBN: 978-1-7374991-1-4 (softcover)
ISBN: 978-1-7378238-1-0 (eBook)
LCCN: 2023906767

First Edition.

Other Books by David Ferrier

THE MOUNTAINTOP SERIES:

Born On A Mountaintop (Book One)

Raised On Rock (Book Two)

Forged In Fire (Book Three)

Wired For Sound (Book Four) - Coming in 2024

California Dreaming (Book Five)   Coming in 2024

After Your Military Service

A Cry For Mercy In The Night

Self Inflicted Wounds

If you enjoy this book, I would really appreciate a brief review. Your help in spreading the word is greatly appreciated and reviews make it easier for readers to find books.

Sing in Me Muse, Of Martial Tunes. When Drums Did Beat, And Children Learned To Listen.

Dedication:

This book is respectfully dedicated to the men of the 571st Medical Attachment (Helicopter Ambulance) with whom I was privileged to serve during my tours in Vietnam.

While this book is, in large part, a work of fiction the characters I have written of are as true as the heroic deeds they performed. To respect their privacy, I have chosen to use fictitious names in my narrative. They know who they are, and I am very proud to be able to write of them.

The men of the 571st Medical Detachment. Phu Bai "Dustoff", San Antonio Texas, many years after. This book is reverently dedicated to them.

"Never think that war, no matter how necessary, or how justified, is not a crime."

Ernest Hemingway

Prologue

Our country's involvement in the Vietnam war began with the collapse of French colonialism and their defeat at Dien Bien Phu in 1954.

The generation Born On A Mountaintop and Raised on Rock would be indelibly altered by this conflict, for and against, until Saigon fell to Communist forces on April 30, 1975.

David, Teddy, and Margaret Mary were engulfed by these events. Running to, or running from their convictions, their lives and their outlooks were forever changed.

These stories chronicle the bloodiest of those times on the war front and the most catastrophic instances on the home front. Many sacrifices were made, cherished illusions were shattered as a generation was Forged in Fire.

Author's Note:

Forged in Fire is not all true, but most of it is, the rest should be.

Table Of Contents

Chapter One

Ferrier, as in Terrier

"You motherfuckers have got exactly thirty seconds to unass this truck and line up on my company street! Now Move! Move! Move! Faster you maggots! The last man off this motherfucking truck is gonna' get my boot so far up his ass he will taste shoe leather!"

And so, we piled out, nobody wanting to be the last man, the one who would find out what this Drill Sergeant's shoe leather tasted like.

It has been three days since I left Lowell. Three days since I raised my hand and took the enlistment oath at Boston Naval Yard. Three days since my very first ride on an airplane. Three days since I had a decent night's sleep. Two days since I had my head shaved bald, every stitch of my clothing replaced with khaki green, and learned to eat as fast as I could off a metal tray with a tin fork, in absolute silence, while somebody called me a motherfucker. Two days ago, I had never heard the word "motherfucker", and Lowell is a pretty blue collar, blue language type of town. Now I head it about every thirty seconds.

Less than one hour ago we left the Fort Jackson, South Carolina Army Reception Center. That is where all the head shaving, clothes swapping, fast eating, getting cursed at business began. For two days we alternately filled out dozens of forms we didn't have time, or permission, to read, had clothing, boots, helmets, tents, ponchos and God knows what else thrown across a counter at us, received inoculations of no explainable origin, taken indecipherable written tests we were not told the results of and been called "maggots", "shitbirds", "assholes", and of course, "motherfuckers" every thirty seconds, or less.

I couldn't remember eating. I couldn't remember sleeping. Just a lot of standing in lines, being screamed at and general terror. I remember thinking, "What the hell did I do to myself?" And "Why are these people so angry? I volunteered to come here, why am I being treated like a convict?" The rest was like images you'd see in a lightning storm, brilliant bright, then black again.

There was the line, at breaking dawn, outside a plain, wooden, one story building we would discover was a barber shop. No talking, eyes front. Four at a time through the door. We exited on the other side of the building. With completely shaved heads and extinct personalities.

Before the barber buzz there were semblances of individuality among us. A Brylcreemed Elvis curl there, crew cut there, semi-Beatle cut there, a bushy Afro or two, quick references on who was who. Now there were only peach-fuzzed skulls, bashful eyes, and uncertain glances, all dressed exactly alike in brand new Army green.

"Drop those bags! Eyes front! You are at attention!" A perpetually angry Drill Sergeant screamed as we leaped out of the truck. He would alternately snarl, growl and swear until the last man was in the street. Despite the threat he did not get a boot up his ass.

We were standing on our Basic Training Company street, a one lane blacktop between parallel rows of two story, run down looking wooden barracks. We were divided into four ranks of ten men each. We were about to start our eight weeks of Basic Combat Training.

"When I give the command to fall out, you will pick up your bags, enter the barracks in front of you, place your duffel bag on a bunk and fall back out into the street. You will stand in the same place in the same line you are standing in now. The first two ranks will select a bunk on the first floor, ranks three and four will select a bunk on the second floor. You have two minutes, maggots, now Fall Out!"

We did. We weren't fast enough. We fell out again. We fell in again. And again. That is pretty much what went on for the next eight weeks. Lots of push ups, sit ups, long runs with lots of gear, threats, promises, and cursing, lots and lots of cursing. Then we were done. There was a parade, sort of. We marched around a parade deck. We graduated. The Drill Instructors who had yelled, screamed, threatened, bullied, bashed and brutalized us for eight weeks smiled and told us we had done well. We smiled back, our chests puffed out with pride.

"Ferry-ay", (that was me), "Get your ass over to the orderly room. The First Sergeant wants to see you, "Move!" Apparently, the yelling wasn't completely over with, at least for me. I entered the orderly room, removed my cap and stood at attention in front of the First Sergeant's desk. He shuffled some papers on his desk. He looked disgusted. He did not look up at me.

"We have not received orders for you Ferrari-er. What MOS were you expecting to be trained for?"

"Helicopter school, sergeant," I answered as confidently as I could.

"Helicopter school?" The First Sergeant was looking up at me now. He did not look happy. "Do you expect the Army to send you to school to become a helicopter?"

I was blushing now, embarrassed. The First Sergeant was turning red too. I don't think he was blushing, or embarrassed. "P-P-Pilot school, Sergeant," I stammered.

P-P-Pilot school! There is no such goddamn thing in my Army as P-P-Pilot school!" Definitely not embarrassed or blushing. He was pissed. And he was staring at me.

"Pilot training Sergeant. For helicopters. That's what I enlisted for."

"Then why didn't you say so in the first place, Furry-er? Did you come in here to waste my time?"

Eight weeks of basic training had taught me when not to answer a question.

"Here is what you will do," First Sergeant Hoadwonick continued, "You will go back to your barracks, and you will remain there until I can find out what happened to your orders for P-P-Pilot school. You will report to this orderly room every morning at 9 AM to see if your orders have arrived. If they have not, and I see you again before 9 AM the next morning you will dig a series of deep and unnecessary holes until sundown. Do I make myself clear, Furnier?"

He did. I got the hell out of the orderly room. The barracks where I lived during my eight weeks of basic training was empty now, except for me. Everyone else had received orders and were off to their advanced training schools. All of the drill sergeants and training cadre who had instructed us were also gone for a two week leave before another herd of basic trainees began their cycle. I looked around the empty squad bay and chose a bunk in the far corner, as far from the door as I could get. Invisible is what I wanted to be. It was not quite going to work out that way. An hour later, as I was sitting on my bunk wondering what to do next another soldier arrived in the barracks. He did not see me at first, skulking in the far corner, but I watched him warily. He was carrying his duffel bag and selecting a bunk to throw it on. Then he spotted me.

"Hey," he began. "Are you a holdover too?"

A what? I wondered. The stranger tossed his bag on a front row bunk and sauntered over to where I was sitting/hiding.

"My names Don Canney." He was holding his hand out to be shaken. I shook it. "So, you a holdover or what?"

"I guess so. I didn't get any orders after basic," I answered.

"Me too, sucks. We're going to be here for a while." Don seemed a lot more up to date than I was. "What did you say your name was?"

I told him. He pronounced it right. The barracks was very quiet.

"So, what exactly, is a holdover?" I asked.

"Means somebody screwed our orders up. We have to stay here until it gets fixed. My DI told me it could take a couple of weeks."

Weeks? I was thinking of a lot of deep and unnecessary holes. And then another guy wandered into the barracks. He looked as lost and clueless as I felt. He froze in the doorway when he spotted us.

"Welcome aboard," Don called out cheerfully. "Pick a bunk, any bunk."

The new guy dropped his duffel bag and looked around. Leaving his bag by the door he walked over to us.

"Hi, I'm Rayell Park." Rayell was as black as a long playing record. Shiny black, built like a bag of muscles. Big eyes, lots of dazzling white teeth. Friendly smile. There were more handshakes. We traded names.

"Anybody know what we are supposed to do now?" Rayell asked.

"Stay out of sight, lay low, wait for our orders to come in," Don answered.

It soon became obvious that Don had been briefed far better than either Rayell or I. His DI, he explained, was a good guy, pro-football player doing his six months active duty. He had played college football against Don before the Army. They recognized each other and Don had a mentor.

"Billy Delmore," Don explained. "They called him 'The Cannon.' Played for the Houston Oilers last year. Now he's Sergeant Delmore, not Billy or The Cannon. He told me this happens to a couple of guys every cycle. Sooner or later our orders will arrive and off we'll go. What school are you guys waiting on?"

"RTO" Rayell answered. "Radio operator, communications school. I was supposed to go with a couple of guys from my platoon. They went, I didn't."

"How about you Dave?" Don asked.

"Pilot training, helicopters. I passed all the tests, I think. No orders though."

"That's Fort Rucker, Alabama," Don explained, "Ten months training. When you graduate, you're a Warrant Officer. Then you still have three years to go on your enlistment."

Somehow this was all news to me. Perhaps I wasn't listening carefully when Sergeant Woods explained it to me but suddenly, I became aware that I had signed up for an extra year in the Army. Four years. This seemed like a very long time.

"You look upstairs yet?" Don asked me. "Cadre rooms, top of the stairs, two of them. They are private, with doors and locks. I suggest we occupy those rooms."

"Won't we get in trouble for doing that?" I asked.

"We are already in trouble," Don explained. "Every asshole who's looking for warm bodies for some kind of work detail is going to check the holdover barracks first. Most of them are too lazy to check upstairs and even if they do, if we are in the cadre rooms, they won't find us. The less we are seen for the next few weeks the better. Get it?"

"What about our orders? Sergeant Hoadwonick said to check with him every morning."

"He's going on leave today. I checked. A Spec 4 named Johnson is going to be running the orderly room. He said if our orders come in, he'll post a copy on the bulletin board after lunch. No need to come around before then."

I would learn a lot about how the Army worked from Don. He received a lot of good advice from his buddy, 'The Cannon,' and Don passed it along to Rayell and I. The three main rules of getting along and getting ahead in the Army are:

1. Don't make trouble for me, I won't make trouble for you.

2. Having no excuse is the best excuse.

3. Forgiveness is easier to obtain than permission.

I would learn these lessons well. Over the next few days Don, Rayell and I figured out where we could eat chow without attracting unwanted attention, where to spend the day without being caught, and how to creep by the orderly room bulletin board after noon chow to see if our orders had come in yet. Thus far they had not.

One night Don and I were tossing a baseball around behind our barracks. Don found a supply of much used sports gear in a closet in the company rec room, which we were not supposed to go into. When we did, we requisitioned two baseball gloves and a ball.

"Canney! You still here?" A compact, brawny, strangely smiling Drill Instructor shouted as he ambled toward us in the twilight. For the past week Don, Raylen and I had become rather skilled in staying out of sight and not attracting attention to ourselves. Now a DI found us. Even more strangely he seemed friendly.

"This is where the Army needs me most, Drill Sergeant," Don replied. They shook hands.

"This is Dave Ferrier," Don said. "Another of Uncle Sam's lost lambs." First time I ever shook hands with a Drill Instructor.

"You guys play much ball?" The Cannon asked. Don nodded, I did too. "Post team is having tryouts tomorrow night. Why don't you come on down, give it a go?" The invitation sounded friendly enough, but Don hesitated.

"Sounds a little high profile, we're trying to lay low, you know?" The Cannon knew, he knew lots of things. "You guys come on down, you make the post team you'll have all the profile you'll ever need. Trust me on this."

So, we did. There were maybe thirty guys on the field when we arrived. The Cannon spotted us and called us over. "Don, Dave, this is Colonel Parker. He coaches the team and plays a mediocre second base." The Cannon and the Colonel exchanged a grin. Then Colonel Parker added, "You should keep in mind however that I can transfer your ass to a very remote icicle station in Greenland for any mediocre number of reasons."

The Cannon smiled, sort of, and we shook hands with Colonel Parker. First time I ever shook hands with a Colonel.

"What positions do you play?" Colonel Parker asked.

Don spoke up, "First base."

"Good," Colonel Parker replied, "we need a bat there."

I looked over the infield where players were taking ground balls. At third base, my preferred position, a Master Sergeant the size of a two-car garage was gobbling grounders and rocketing them over to first base. Second base, my back-up choice, was currently being played by a mediocre Lieutenant Colonel who was waiting for my answer.

"Outfield," I answered, "Either corner." There were about ten guys out there shagging flies. I figured I'd either fit in or regain my anonymity with that group.

"All right then, go warm up. Sergeant Cannon, a word."

A couple of hours later Colonel Parker called us into a circle. Several guys were thanked for coming, the rest of us were told we were on the team. Practice every Tuesday and Thursday evening, game days were Saturday afternoon with an occasional Sunday double header. Then Colonel Parker explained the rules.

"Once we cross those base lines, we are a team. There is no rank, we play as equals, teammates. On the field you may address me by my nickname, which is "Sir", the rest of you can decide what to call each other as you see fit. I prefer to win and we will try our best at all times to do so. Should we lose we will be as gracious in defeat as we are in victory. Playing on this team is a privilege and any behavior which reflects poorly on the United States Army, or yourself, will not be tolerated. Sergeant Cannon is your team captain. Direct all your on-field questions to him. Let's play hard, have some fun and beat the bastards as often as we can."

It was a good speech and playing on the team was the most fun I'd had on a ball field in years. The ticking bomb was the looming reassignment orders Don and I were waiting on. Rayell had shipped out a few days after the season started, here one day, gone the next. Then one afternoon my orders arrived. Fort Rucker, Alabama. Pilot training.

"You cannot depart this post until after our season ends," Colonel Parker announced. "We are playing Fort Benning this week and Fort Gordon the week after. If we win those two games, we are going to the Regional Finals at Fort Sam Houston. You are therefore essential personnel. Pilot training is going to have to wait."

Sure enough I was placed on a "medical hold", whatever that was, and finished the season. I learned a lot while playing ball that summer. First and foremost was that the men I served with, who were serving their country, were a really good bunch of guys. Black, white, Colonel, Sergeant or Private it seemed you were treated as decently as you treated others, as decently as you behaved. I also learned that I was a pretty good softball player. I could "hit 'em where they ain't", and run down a lot of big, floating fly balls with the best of them. I played left field or right field, depending on which side of the plate the other team's power hitters were, and I punched a lot of base hits into right field. Most guys, when they hit a softball, give it the big, uppercut swing, which sends a lot of lazy fly balls out to a lot of energetic outfielders. I would swing level, keep the ball on the ground and between infielders. I learned to switch hit and could reach the fences from either side of the plate. Life was good. Don, it turned out, was an excellent first baseman, great range and power. We won a lot of games that summer and went to Fort Sam Houston and won and then to Fort Ord, California where we won as well.

When I wasn't playing softball, I was something called a "barracks orderly" which went along with my medical hold, which held off any pesky reassignment orders. Like pilot training and an extra year in the Army. Softball and barracks orderly worked well until the day First Sergeant Hoadwonick summoned me to his office.

"Get in here, Ferrara-ay!" Sergeant Hoadwonick shouted from behind his desk. I got in there as quickly as I could. "Are you trying to rat-fuck my Army, Fornier?"

"No Sergeant," I stammered. I wasn't sure I would know how to rat-fuck Sergeant Hoadwonick's army if I wanted to.

"Then how do you explain this medical hold bullshit! You don't look like you're on any medical hold to me!"

He had a point there. Then I gave him the undoubtedly worst answer I could have. "I've been playing softball, Sergeant."

"Softball!" I thought he was going to come across the desk at me. "What in the name of the bleeding, humping, limping Jesus does softball have to do with a medical hold?"

This was another of those questions I was better off not answering. I stared stoically at a spot one inch above Sergeant Hoadwonick's head and waited. He angrily shuffled several papers around his desk before pushing one sheet toward me.

"You have orders to report to Fort Meade, Maryland for further assignment to APO 96308. These are overseas assignment orders. They supersede any medical hold. APO 96308 is in the Republic of South Vietnam. Playing softball is over. You have one hour to clear this post and report back to me. Now get the hell out of my office!"

I did. I had no idea how to "clear post" so I went to find Don. He was no longer a barracks orderly. He was now assigned as permanent party at Fort Jackson.

"This is some serious shit." Don proclaimed.

"What am I supposed to do?" I asked. Don didn't know. Sergeant Delmore was no longer on active duty. We decided to ask Lieutenant Colonel Parker.

"You get your ass up to Fort Meade and report in. Explain to them you have received no advanced individual training. You should then be reassigned to a training unit."

Colonel Parker was quite clear in his instruction on how I should clear post. I picked up my finance records and personnel file and hurried back to Sergeant Hoadwonick's office.

"Get in here Fer-ary-aria! Do you have your records?"

I waved the large brown envelope I had been given at him. He tossed my reassignment.

Orders to me. "Now get the hell out of my office!"

"Could I ask one question before I go, First Sergeant?" I asked as respectfully as I could.

Sergeant Hoadwonick stopped shuffling papers and leaned back in his chair. He looked impatient and curious, as well as angry, as usual. He nodded.

"Are you familiar with a small, short haired, wiry dog called a Terrier, Sergeant?"

Sergeant Hoadwonick looked puzzled, impatient, curious and angry, as usual. He nodded.

"It's Ferrier, as in Terrier." And I got the hell out of his office.

Chapter Two

The Green, Green Grass of Home

Before I had to report to Fort Meade, I was granted a one week leave to my home. I rode a Greyhound bus from Columbia, South Carolina to Lowell, changing buses in Washington, D.C. and New York City. I thoroughly enjoyed the long bus ride, for the first time in the six and half months I was alone. Alone on the bus, alone in the bus terminals, able to think my own thoughts in my own time. And I had a lot to think about.

The Army was starting to feel more like home than my destination. I was comfortable there having learned the do's and don'ts and how to go along and get along. I was promoted to Private First Class along with Don, midway through softball season by Colonel Parker. I learned the privileges and status of rank, any rank, and wanted more. So far, the service had been an adventure and truly a life changing event. I felt more independent, more grown up, even though the Army was the most structured and rigid environment I ever encountered.

After an initial wave of homesickness, I settled in well to military life. I wrote home during boot camp, of course, mostly just to say I was alright and to let my family know I missed them. But as time went on, the missing diminished, my sense of belonging increased and my self-confidence flourished.

On the night I arrived home my parents picked me up at the bus station downtown. Though I had only been gone six months they looked older. Downtown Lowell looked smaller. I felt out of place and did not want this to show.

That night sleeping in my childhood bedroom I dreamed of walking next door to see Teddy, or over to see Margaret Mary. In the cold light of day, neither would be there. Teddy was immersed in the Mean Green Machine known as the United States Marine Corps. Margaret Mary was in Illinois, in college, away from me, out of my life.

There was little doubt that Teddy and I were on the fast track to Vietnam and whatever fate awaited us there, but right now, in the early morning fantasy of my bed I wanted to go next door, get Teddy and go for a ride on our bikes. I wanted to hear Margaret May laugh and hold my hand. Those days however, felt long past and never returning.

That morning, after my Mother made my favorite breakfast of French toast and link sausages, I put on my khaki uniform and went next door to ask about Teddy. Connie answered the door with a big smile and a hug. She asked about me. I asked about Teddy. He was in Okinawa by his latest mail. Connie retrieved his letter. Teddy wrote…

Hi Dad & Connie,

I am still on Okinawa but I don't know for how long. We packed our sea bags yesterday and are on movement alert. I guess you know that means Vietnam but don't worry. I am ready and so are my buddies. The Marines are the toughest fighters in the world and I am one of them Connie, how do you like being married? Is Nick being good to you?

He better be because you deserve it. Has anyone heard from Chris? Is he still in the hospital? Please let me know as soon as you find out. How about Dave? Is he done with his training yet? Ask him to write to me when you find out okay? You can write to this address because even if we are gone the letters will get forwarded to us. I miss all of you and think about you a lot. Dad don't work so much if you don't have to. I have to go now, or I won't get chow. I love you all,

Your brother and son and proud Marine, Ted

Connie wrote down Teddy's address and I left to go over to Margaret Mary's house. There was a "For Sale" sign on the lawn. My Mother explained that Sean had moved to Illinois and was living with his sister. He had been offered his teaching position back at Loyola and accepted. He left me a note which my Mother gave me. Sean wrote…

David,

I hope this message finds you well. I accepted a position at Loyola in the hope that when my blessed daughter returns to her education I shall be there for her. I hope you are faring well in your military career and that you may stay safe until we meet again. I have left my address with your loving Mother, please stay in touch with us so that we may stay in touch with you,
Slan Leat,
Sean

And another piece of my childhood crumbled to dust.

The rest of the week I wandered around my hometown like a ghost. An occasional familiar face said a vague hello. I went by Lefty's; they were glad to see me but very busy. Shedd Park was semi-deserted except for a few young kids. Most everyone I knew was gone. I went by the gym. Max was happy to see me and introduced me to a roomful of fighters I didn't know. Max wished me well. I promised to write.

One highlight of my leave was a visit to see Sergeant Lewis. I showed up at his office decked out in my full military regalia, crisp khakis, National Defense Ribbon, Expert Rifleman's Badge and Private First-Class chevrons. I was stylin', he was smilin'. "Private Ferrier," he exclaimed when I walked into his office and snapped to attention. "I thought you would be a Peter Pilot by now."

I told him all my Fort Jackson war stories. He told me medical hold was hardly ever a good idea, especially when you are awaiting orders for flight school. He also told me I could have taken a plane home, military standby, because I had transfer orders. He fixed it up so that I could fly back to Baltimore to get to Fort Meade. No more Greyhound bus for me.

"Don't try and game the system, Dave. The Army rewards effort, not ease. Sometimes the hard way is the best way. You'll learn. Keep learning. I'll try and locate those cancelled orders for you. Call me when you get to your duty station and I'll tell you what I found out."

Good guy, straight shooter, always.

On the day before I was scheduled to leave I went to Paradise Donuts with my father. We sat at a familiar corner table and had a very unfamiliar talk.

"Dad, the unit I'm reporting to is forming up to go to Vietnam. I don't know whether I'll be going with them or not."

"Why would your unit be going and not you?" He asked.

I told him about the whole softball thing and cancelling my orders to pilot training. He knew, of course, that I had been playing softball, what he didn't know is that I hadn't been doing much of anything else.

"I'm supposed to explain all that when I get to Fort Meade. I don't know what's going to happen after that."

My Dad thought for a moment. Sipped his coffee, took a deep breath. "Get as much training as you can, son. If you have to go into combat, go with as much training as you can get."

My Dad would know about going into combat. He had flown with the Eighth Air Force as a tail gunner on B-24 bombers out of England in WWII. He had never talked about it with me before, maybe we would talk about it now.

"Are you still interested in becoming a pilot?" He asked.

"Sort of. I don't know if they will let me because my orders got cancelled."

"So, you could play softball?"

"Sounds kinda' ridiculous now, doesn't it?" Was all I could say.

"Most times life is a combination of ridiculous and reasonable, son. Choose reasonable whenever you can."

Which was one of the coolest things my Father ever told me.

"David, there's something I need to tell you about going into combat." My Dad leaned forward across the table and looked me right in the eyes, "You are going to be frightened, scared senseless, if you have any sense at all, which I know you do."

This was a long speech for my Father, and it was about something I always wanted to talk to him about. Somehow, he seemed to know that now was the time I needed to hear this, not merely wanted to hear it.

"Try to remember, being brave is not about not being scared, it's about doing your job even though you are scared. That's where training comes in, son. Get as much as you can, pay attention, and be with other people who are doing the same thing."

"Is that what you did when you were a tail gunner?" I asked.

"It's exactly what I did. Be with the people who are doing their job well. Heroes aren't only the ones who get the medals, heroes are the guys who do their job no matter how scared they are. Be like them and you will be a hero too."

I couldn't think of what to say back. My eyes had filled with tears, and I had a giant lump in my throat. My Dad's eyes looked kind of teary too. Finally, I croaked, "I will, Dad, I promise."

"Good," My Father straightened up in his chair and smiled, "And write to your Mother. She worries about you."

That evening we had a very special going away dinner at Glenmere Street. My Mother made all my favorite foods and there was a cake, chocolate, of course.

I could feel the sense of foreboding in the room. My brothers were much more quiet than usual. My Mother was anxiously cheerful though I saw her hands shake when she passed me a plate. My Dad and I had made our peace over donuts. I now tried to do the same with the rest of my family over the cake.

"Thanks Mom, for this delicious meal," I began, "and for all the meals actually. I really do appreciate it and I miss your cooking a lot. There is a whole lot I've come to like about the Army but the good food, for me, will always be right here, with you guys."

If I could have thought of something to say to lighten the mood I would have but couldn't. So I continued, "I don't know where I'm going or what I'm going to do when I go back, but the unit I'm assigned to is going to Vietnam and if I have to, I'll go with them."

That did very little to lighten the mood but at least the topic was on the table now.

"I know Vietnam is dangerous and if I go there, I want you to know I'll be careful and do the best I can to come home to all of you. If I don't and something happens to me, I need you all to know that I love you a lot and always will."

There were some tears then, a lot of hugging and smiling and of course, there was the cake.

The next day my family drove me to Logan airport. This time our goodbyes were more heartfelt and our parting more solemn. From the airplane window I could see them in the terminal waving at the airplane even though they could no longer see me.

I realized that as I travelled into the next part of my personal adventure, I would miss more fondly than I ever imagined what I was leaving behind.

Onward.

****"Slan Leat" is a Gaelic farewell, literally, "Goodbye, health be with you."

Chapter Three

Off To See The Wizard

My flight from Logan Airport to Baltimore was uneventful save the wrenching of my emotions, which like some invisible elastic band was stretching thinner and thinner as I flew on, finally snapping as we landed at Friendship Airport. I was back in the Army now. The next phase of my adventure was about to begin.

After showing several pairs of roaming MPs my orders, I was able to find a military shuttle to Fort Meade, Maryland. That's when things started to go off the rails. Upon arriving at the Fort I was told to report to the Post Locator, which is where the shuttle dropped me. So far so good. Inside I handed my orders to a very bored Specialist Fourth Class. My orders clearly stated that I was assigned to the 571st Medical Detachment (HA), which I found out meant Helicopter Ambulance. The Specialist Fourth Class then discovered there was no 571st Medical Detachment at Fort Meade. Nothing even close. Worse than that he was looking at me like I should know what to do next. When he finally concluded I did not, he dragged himself across his office to the desk of his supervisor, a more experienced, and more helpful Staff Sergeant, named Whitcomb.

"We seem to have a bit of a dilemma here Private," Sergeant Whitcomb announced.

At least he was saying "we". "There's not much I can do about this today, but let's get you a bunk for the night and I'll check on this tomorrow."

So far my military career consisted of no orders, cancelled orders and orders to nowhere. But they did find me a bunk in a transient barracks and told me to come back tomorrow. The rest was pretty easy to figure out. I ate chow, found the base movie theatre and saw a movie, slept in, had breakfast and went back to see Sergeant Whitcomb.

"I've got some good news and some other news," Sergeant Whitcomb announced. "There is a 571st Medical Detachment, that's the good news, the other news is you are it, at least for now. Come on, let's take a ride."

On the ride to a far corner of the base Sergeant Whitcomb explained that the 571st Medical Detachment was forming here for further deployment overseas. Overseas meaning Vietnam. I was the first man to arrive. He had no idea when others would follow other than "pretty soon". He parked the jeep in front of a standard, two story, wooden barracks. The bottom floor was occupied by a cheerful herd of clerk typists in something called an "AFDOJ" Admin center. I never figured out exactly what that meant. Sergeant Whitcomb told me to follow him upstairs. There was a large open area with a desk, a swivel chair, a telephone and two large metal file cabinets. There were two smaller offices against the far wall with nothing in them.

"Welcome to the 571st Medical Detachment, Private Ferrier. As of now you may consider yourself acting First Sergeant, Supply Sergeant, Company clerk, and chief cook and bottle washer."

Yesterday I was a high private with no Military Occupational Specialty and no advanced training other than barracks orderly and outfielder. Now I could add bottle washer, and all those other things.

"Relax," Sergeant Whitcomb chuckled, "All you need to do is answer the phone, which isn't hooked up yet, file the daily distribution which you will start getting today, receive and sort the mail, fill out a morning report every morning and sign in any additional personnel I send over to you."

"Sign them in how?" Was all I could think of to ask.

"Come on, I'll introduce you to Kearney, he'll fix you up with everything you need."

We went downstairs and I met Specialist Fifth Class Ray Kearney. He was a good natured, red-headed Irishman and head of the personnel section on the floor below me. Sergeant Whitcomb drove off with a smile and Ray Kearney held out his hand. We shook, he led me inside.

"You're gonna" love it here. Nobody bothers us, we get our work done, we close up for the day. Strictly nine to five, with generous lunch and coffee breaks as needed. Let's get you some gear."

Specialist Kearney went to a bank of file cabinets and began handing me stacks of forms. White forms, blue forms, pink forms, yellow forms, perforated forms, forms with lots of carbon copies attached, long ones, short ones and a green ledger book.

"That oughta' get you started. Can you type?" He asked.

"I can spell." I didn't want to oversell myself.

"Close enough. I'll get your phone hooked up this afternoon. Any questions?"

I had a few. Within a day or two I was answering the phone like a champ. It hardly ever rang and when it did it was usually a wrong number. I did learn to answer by saying "571st Medical Detachment Orderly Room, Private Ferrier speaking." The wrong numbers hung up then. I took careful messages for the rest and stacked the pink message sheets in a desk drawer. I found out that "distribution" meant a large overstuffed brown envelope full of mimeographed announcements and schedules and the like which came every morning around 10 AM. I stacked those next to the pink slips. The rest of the time I swiveled in my swivel chair, read paperback novels and closed up whenever Ray Kearney and the guys downstairs left for the day. This rather pleasant arrangement lasted about a week, then one evening as I stood in line at the mess hall I heard a voice ask, "Anybody know where I can find the 571st Medical Detachment?" This was not just any voice, however. This was the thickest, twangiest, hill-billiest accent I ever heard. I looked around to find a large, friendly-bear-like individual with a baffled look on his face. Without thinking it through clearly, I said, "I'm the 571st Medical Detachment." And now we were two. That voice belonged to Private Larry Woodman, a native of Richwood, West, By God, Virginia who was now assigned to the 571st Medical Detachment. "Woody" had been in the Army less time than I had, but he had been to a training school after basic. He was an Avionics Repairman. I wasn't sure what that meant. Radios, he explained, radios that went in helicopters. A picture was starting to form.

Woody and I ate chow that evening and swapped life stories. Admittedly these were a bit brief at the time, grade school, high school, a faint whiff of college. Woody was a star athlete at Richwood High, football was his main claim to fame, but he also played basketball and baseball. We traded sports stories, girlfriend stories and basic training stories. Woody had enlisted, in part, to get enough money to go to college after his discharge. That sounded like a better idea than mine, so I made a similar claim. It became obvious that Woody was a good guy, honest, dependable, capable and willing. Our friendship began.

After chow I got him squared away with a bunk in the empty barracks I was staying in and even got him properly signed into the unit roster, thanks to Ray Kearney.

"What do we do now, Dave?" Woody asked the next day in the orderly room.

"We wait. I'll let you know when it's your turn to answer the phone."

Two days later three guys arrived. They were all privates, right out of medic training at Fort Sam Houston Texas. Sergeant Whitcomb dropped them off downstairs. They were: Harry Giles, Mario Calini, and Dwight Anderson. This was their first duty assignment. I signed them in, Woody found them bunks and now we were five.

The next day a guy wandered into the orderly room. He was wearing madras Bermuda shorts, a gray University of Virginia t-shirt and flip-flops. He had a congenial enough smile and asked, "Who's in charge around here?"

"Nobody," I answered truthfully. "Can I help you?"

"I'm a pilot. Name's Bruce Tyler. Am I supposed to sign in or something?"

I signed Bruce Tyler in, discovered he was a Warrant Officer, right out of flight school. This was his first assignment as well. I explained the situation to him, he explained his situation to me, "Look," he said, "I'm from Richmond, about an hour from here. I'm going home to spend some time with my family. Call me if anything important happens." He wrote down his phone number, jumped back in his sports car and left. That made us six.

That afternoon I got a phone call from some angry sounding sergeant wanted us to come down and "Get our goddamned vehicles out of his motor pool before he gave them to someone else."

I learned there were four vehicles, two jeeps, a three-quarter ton truck and a deuce and a half screwing up this guy's parking lot and that we should come get them right away. I rounded up Woody, Harry and Dwight and off we went.

"Sign here," the unhappy sergeant demanded when we arrived. I wondered how he got by in the Army without ever shaving and signed some form I'd never seen before. He gave me a copy and off we went. The 571st Medical Detachment had wheels. This was going to be fun.

After we parked our brand-new vehicles in front of the orderly room, I put the form I signed in the drawer with all the messages and distribution and tried to think of an excuse to go drive around the post. Woody and Dwight took off for parts unknown, but Harry stayed around. He wanted to drive around the post too.

Harry Giles was an eighteen-year-old medic from Philadelphia. He was married to his high school sweetheart, Linda. He had a baby daughter, named Cindy. They lived with Harry's parents in Philly. He spoke of his wife and child fondly, sometimes referring to Linda as "Belinda", which turned out to be far more interesting than a mere oversight. Harry was much like having a barely housebroken puppy around. He had an unquenchable good humor, very little common sense, and boundless, usually misdirected frantic energy. He seemed unable to start a sentence of any kind without using the word "Yo" first and often after that.

"Yo Dave, yo, let's go for a ride in the jeep. We can see what's going on around here." Harry moved a lot when he talked. Harry moved a lot when he wasn't talking.

"Can't right now Harry," I said, "Somebody's got to be here in case the phone rings or more guys show up. I'll tell you what though," I remembered one of the pink slip messages I had received the day before and rifled through the pile in the drawer till I found it. The message said that the following items from our TO&E could be picked up at Building S-4. Then there were a lot of numbers. I handed Harry the message.

"I have no idea what this means. Go ask Ray Kearney and then go pick this stuff up, whatever it is. Okay?"

Harry bounded off like a puppy with a new pull toy. I swiveled in my swivel chair. The phone did not ring. Three hours later Harry came back. He was all smiles.

"Yo Dave, yo, there's this whole building full of stuff over there. Some sergeant said we have to come pick it up."

Thirty minutes later Harry and I were being chewed out by Master Sergeant James Hendricks. What I could make out between curses was that our TO&E was screwing up his TO&E and we should get ours out of there immediately.

"Where am I supposed to take this stuff Sergeant?" I asked.

"How the hell do I know where you're supposed to take it! Just get it the hell out of here! Now! Today!"

I remembered Rule Number One about getting along in the Army. Apparently, I was making trouble for Sergeant Hendricks, or at least all this equipment was. Rule Number Two was not to make excuses, so I told him I'd get right on it. Rule Number Three was the easier to get forgiveness than permission clause, so I figured I'd stack what I could in the orderly room and figure out what to do with the rest later. When Harry and I got back to the orderly room there were two guys waiting there, Private First-Class Johnnie Kane and a Specialist Fifth Class. I was very happy to see the Spec Five, he was a sergeant, let him be in charge for a while.

"I'm not gonna' be in charge of shit!" The new Spec Five, named Dennis Hughes, proclaimed, "I'm a crew chief. I fix helicopters, I fly in helicopters. That's what I do and that's all I do. Get somebody else to be in charge. Where's our barracks?"

The new guys left with Harry to draw bedding and unpack at our barracks. I was back in charge again. I told Harry to round up whoever he could find and bring them back to the orderly room. Harry came back with Woody and Dwight, the two new guys and another guy I had never seen before. His name was Lewis, he was looking for the 571st, got lost, found Harry and came along. I signed him in, and we all loaded aboard the deuce and half and three quarter trucks and set off for S-4. When we got there Sergeant Hendricks started shoving papers at me and telling me to sign them. Before I did, Woody, bless his heart, taught me Rule Number Four about getting along in the Army. "Count 'em before you sign for them."

"Says here we're supposed to have eight fox mike radios and I only got four," Woody said.

Sergeant Hendricks muttered something about the other four probably got put "somewhere else". By the end of the day, we found that quite a few items had been put "somewhere else" but we found them and loaded all of it onto trucks.

I never signed so many forms in my life and I read every one. Sergeant Hendricks got grumpier as the day went on, especially when we found the misplaced items, but when we were done, he told us to be careful when we got to 'Nam' and always count our chickens whether they were hatched or not. Over the next few days several more guys showed up. Four of them were Warrant Officers right out of flight school. They wanted to know where the helicopters were parked. The rest were a mixture of privates, a Spec Four or two but nobody I could talk into taking over in the orderly room. There were no sergeants as yet and Hughes was still the only E-5 in the unit.

He had become buddies with Woody and one night after chow the three of us went to the Enlisted Men's club for a beer. Hughes got to talking and I learned this was his second deployment to Vietnam. He was there in 1966. When he got home his fiancée had broken up with him, so he volunteered to go back to Vietnam. He might have been having second thoughts about that decision.

"There ain't nothin' but one thing good about bein' over there," he explained, "It's hot, stinking hot, it's dirty, it's dangerous, the food sucks and the water smells bad and there's about a million gooks smilin' at you one minute and trying to kill you the next."

"What's the good thing?" Woody asked as I wondered.

"Dustoff", Dennis replied immediately, "Med evac. That's us. That's why I volunteered to go back. Dustoff's the only job over there getting' done right."

First time I ever heard the term, "Dustoff". Dennis explained that was the radio call sign for a medical evacuation. Front line, hot or not, to get the wounded out. I had no idea up to that moment what the 571st Medical Detachment was all about but from that moment I wanted to be part of it. Two or three beers later we ambled back to the barracks and turned in. The next morning, I went to breakfast with the guys and went to open the orderly room. I asked Woody to hang around the barracks and keep an eye on all the equipment we had stored there when we ran out of space in the orderly room. I didn't know what half of that stuff was, but I had signed for all of it.

I was midway through a routine morning of waiting for the phone to ring when the world turned, the starting bell rang, recess ended and the 571st Medical Detachment became an honest-to-God, military organization. Up the stairs and into the orderly room came a Major, a kind of unhappy-looking Major. He stopped when he saw the stacks of boxes and piles of gear mounded about the room. He sighed, looked at me and asked, "Who's in charge here, Private?"

"Nobody, sir." Which was, of course, the wrong answer.

He sighed again and asked, "Then who are you and what are you doing here?"

I finally began to grasp the seriousness of the situation and sprang to my feet and saluted, "Private First Class Ferrier, sir. I'm kind of keeping an eye on the place."

This time the sigh was a bit deeper. "You do not have to salute me Private. We are indoors. Where is the First Sergeant?"

"We don't have one sir."

"The company clerk?"

"We don't have one sir."

"Where are the rest of the men?"

"They're uh, they're around sir. They generally show up around chow time sir."

"Show up?" He muttered as he looked around the office. He stopped at one of the small side offices and looked in. The room was full of folded cots, tents, tarpaulins and miscellaneous gear.

"Is this my office?" The major asked.

"I don't know who you are sir. Is that supposed to be your office?"

He answered evenly, "I am Major Vincent Cronin, and as of this moment I am the Commanding Officer of this unit. Get this office cleared out immediately. Find me a desk, a chair and round up the rest of the men, Right Now!"

I raced downstairs, found Ray Kearney. He found me a desk and chair. I found the guys, most of them. I hadn't seen the Warrant Officers in two days. I remembered to call Bruce Tyler and told him we had a Major. He said he'd be there in an hour.

Woody, Dwight and I carried the desk up the stairs into what was now Major Cronin's office. All the gear that had been stashed in there was spread around the orderly room.

Major Cronin thanked me for getting the desk and asked me to have the men line up downstairs. We were going to have our first unit formation.

We were a motley crew gathered in front of the orderly room. There were eleven of us all told, not counting the four Warrant Officers I hadn't seen in two days and Warrant Officer Bruce Tyler who was speeding our way in his sports car. Major Cronin took his place in the front of our formation. He called us to attention, placed us at parade rest and said, in a loud and clear voice,

"As most of you know by now, I am Major Vincent Cronin, your commanding officer. We are the 571st Medical Detachment, Helicopter Ambulance. That means we are Dustoff, gentlemen. The primary duty of a Dustoff unit is to save lives, and in a very short amount of time we are going to Vietnam to fly Dustoff missions."

He let that sink in for an increasingly solemn moment. Then he continued,

"Any man who cannot fully commit to this mission should let me know before we depart. Our job is going to be too dangerous and too important to not have people who are giving it their one hundred per cent effort. If any of you feel you cannot do this, I will do my best to have you transferred before we depart. For the rest of you our work is going to be hard, it is going to be dangerous, but it is also going to be the best thing you have ever done in your life. Who is the ranking man here?"

We all kind of looked at each other until Specialist Fifth Class Dennis Hughes reluctantly stepped forward.

"Good," Major Cronin said, "You are now my acting first sergeant. You will see that the men are ready for duty every morning and here for a formation at 0800 hours from this day forward. Do you understand that?"

Dennis saluted and said, "Yes sir."

"Does anyone here have any experience running an orderly room?" Our Major asked next. Nobody responded.

"Private Ferrier, you are now my acting company clerk. The rest of you will be assigned duties as they occur. Sergeant Hughes, I will inspect the unit barracks at 1500 hours today. You will see that it is ready for inspection.

"Sergeant Hughes, dismiss the unit. Private Ferrier, you are with me."

Major Cronin and I went back to the orderly room. He told me to follow him into his office. I stood at attention in front of his desk until he said, "At ease, Private." And then in a less formal tone asked, "What is your MOS, Private Ferrier. I can't seem to find one in your 201 file."

So, I told him the whole sordid, softball, barracks orderly, cancelled orders saga. He sighed, I was getting used to seeing him do that, and said, "I'll see what I can do about straightening that out. In the meantime, you are my company clerk. Now why is all this equipment in the orderly room? Why isn't it packed in our Conexes?"

I then had to admit I didn't know what a Conex was and even worse, where our Conexes might be.

"You should know," he replied, "You signed for ten of them, if you don't know where they are the Army is going to want you to buy ten more."

He did not seem to be kidding. I was thinking about a whole drawer full of equipment I signed for and didn't exactly know where or even what I was signing for. I could hardly wait to tell him about that. Before I could respond, we were interrupted by the arrival of Warrant Officer Bruce Tyler. He was properly attired in crisp fatigues, bright, new Warrant officer insignia and flight wings.

"Warrant Officer Bruce Tyler reporting sir," He said standing at attention in the doorway of Major Cronin's office.

Major Cronin motioned him in and me out. I felt like I had escaped, but from what and for how long I couldn't imagine. As soon as I cleared Major Cronin's office, I took off to find Ray Kearney downstairs. I had to find out what Conexes were and where they most likely would be. Ray explained that Conexes were large metal shipping containers and told me to try the shipping yards by the post railhead. I hurried back upstairs in time to see Bruce Tyler exit Major Cronin's office. Bruce told me he was now acting executive officer, whatever that was, and he was supposed to go find the other four Warrant Officers I hadn't seen since they signed into the unit. He drove off in one of our jeeps to the airfield and Major Cronin called me back into his office.

"Private Ferrier, we have an enormous amount of work to do and a very short amount of time to do it. I am going to try and find a replacement company clerk, and have you reassigned to get some training, but for now I need you to work with me to make some sense out of this mess."

The Major seemed appropriately exasperated by the condition of the unit. I felt responsible, though I couldn't figure what else I could have done. As cheerfully as I dared, I said, "Major, sir, I think I know where our Conexes are."

Major Cronin looked at me, sighed as I explained, and said, "Let's take a ride in one of your jeeps, Private." We found the Conexes, arranged to start having our equipment delivered there and headed back to the orderly room. Major Cronin's mood appeared to have lightened. I felt like I had accomplished something useful.

"Good work, Private," He said as I drove along. I felt like I had been promoted to General.

Over the next three weeks events moved very quickly. Major Cronin was constantly on the telephone making things happen. More men arrived. A Captain, really nice guy, replaced Bruce Tyler as Executive Officer. We got a First Sergeant, an E-7 named Willie Johnson. It took me about three days to figure out he didn't know what the hell he was doing but he was really good at sucking up to Major Cronin, who was too busy to notice how incompetent this guy was. Meanwhile I was figuring out this orderly room business. I learned how to file the forms, fill out the paperwork, get the mail, organize the clutter and keep things moving in the orderly room. I made friends on the base, knew who to call when I had a question or needed something. I was in the mix, part of the mix. Life was good.

Sergeant Johnson stayed out of the way and let me run things. By now we had a full unit, fourteen officers, all pilots, raring to fly, and thirty-two enlisted men, sorted, and as ready as they would ever be to depart for Vietnam.

All of which led to me being called into Major Cronin's office a week before we were scheduled to deploy to Vietnam. "Young man," he began, Major Cronin started calling me young man as our relationship solidified over the past weeks. Nobody in the unit worked harder than he did, and I tried to work just as hard as him. "Young man, I have managed to get you reassigned to a training unit. It's not Fort Rucker, but it could be once your original orders are found."

Reassigned? I didn't want to be reassigned. I wanted to stay with the unit. Stay with my buddies Woody and Harry and Johnnie Kane and Denny Hughes. I realized Major Cronin was trying to do me a kindness, but I didn't want to be sent away.

"I can't take you along without your having an MOS that's part of the unit Table of Operations. You don't have any MOS at all," He declared, with some disappointment.

"I can fix that!" I replied without really knowing how. "I can have a duty MOS in my personnel file by this afternoon. Then can I stay with the unit?"

Major Cronin seemed amused. I think he wanted to see how this could be done as much as I did. "If you can be back in my office by two o'clock with an MOS assigned to your file that fills this unit's vacancy for a company clerk you can go with us."

I was dismissed. I scrambled down the stairs and burst into the lower offices shouting, "Ray! I need your help." An hour later I was a 71H20, a Personnel Records Specialist, same as Ray Kearney. At least I would be if Major Cronin would sign my on-the-job training certification. He did. I was.

Seven days later the 571st Medical Detachment along with me were on our way to Vietnam.

"It is a proud privilege to be a soldier – a good soldier – with discipline, self-respect, pride in his unit and his country, a high sense of duty and obligation to his comrades and to his superiors, and a self-confidence born of demonstrated ability."

George S. Patton, Jr.

"Always trust those searching for the truth, never those who have found it."

-Jordan Maxwell

Chapter Four

The Turning Wheel

"If you weren't so fucking stupid maybe you could make some of this shit we could sell and make some money!" Kendall smashed the clay pot Margaret Mary had been hand painting flowers on all morning. He raised his hand to slap her but she cowered so far back in her chair that he laughed instead and stomped toward the door. Before he left, he turned on her and said, "I owe you one, bitch."

Margaret Mary sobbed into her freezing, chapped hands. Her fingers were swollen from the cold, a large bruise blossomed under her left eye. She tried to pull the thin blanket tighter over her shoulders and looked out the filthy window at yet another foot of early winter snow which had fallen outside. The verdant, green valley they had visited in early summer was now under three feet of snow, cut off from the outside world. A six mile long, unpaved mountain road led up to the valley and it would be impassable until there was more snow melt. Realizing this her tears came harder, and her hope of escaping disappeared.

Six months earlier Margaret Mary and Kendall arrived at the High Meadow Commune of Peace and Love, a conglomeration of college dropouts, runaways, transients and minor league drug dealers. The poster would not have read that way. Kendall described the gathering as a communion of free spirits, intellectuals and enlightened ones who rejected the crass commercialism of society and leaders in the anti-war, anti-capitalism, anti-everything hippie movement. In the spring and summer there were campfires, sing-alongs, peace, love and dope. Lots of dope. Marijuana, grown in secluded patches and cultivated by the enlightened, dominated the day, evening and night activities in the commune. Margaret Mary initially found the drug relaxing, convivial, a harmless alternative to the evils of alcohol. Over the weeks, and then months. she realized that the intellectuals she had hoped to encounter turned out to be lazy, ill-informed, dirty and promiscuous.

Kendall settled right into the new normal. He preached awareness though they were effectively cut off from the outside world, high in the Colorado foothills. When not stoned, he would pontificate about issues he knew little of, impressing the congregation with vocabulary rather than wisdom. He broached, then insisted that he enjoy an "open relationship" with the other females of the tribe and tried unsuccessfully to convince Margaret Mary she should diversify as well. The first time Kendall hit Margaret Mary was when she found him in their bed, sleeping bag actually, with another of the community's females. After that she tried to walk into town. Kendall caught her and dragged her back.

When the summer turned to fall and the weather grew colder the colorful teepees, lean-to's and tents became less organically correct and more icy and damp. When she told Kendall she wanted to leave he snarled at her and forbade her to leave the encampment. The only road out was six miles of forest path. The only vehicles a broken down Volkswagon bus and a jeep with three tires.

To feed themselves these rejects rode the unreliable VW into the nearest town and dumpster dove behind supermarkets for discarded produce and semi-spoiled meat. Margaret Mary was not allowed to go on these excursions and was, by order of Kendall, confined to the camp.

There she found little solace from her fellow outcasts. They insisted she was the root cause of her troubles and increasingly shunned her. There were no phones, no mail, the authorities left them alone and the isolation was complete.

Then the snows came. It was going to be a long winter.

"Put your '60 here," Lieutenant Cross ordered. Traverse from the firing stake on your left at 150 Yards to the Commo Tower. Anything moves out there, kill it. There are no friendlies in the area, and we are effectively surrounded."

"Can do sir," Teddy, now Lance Corporal Gianoulas, responded.

"I'll get you some more men up here as soon as I can, Corporal. In the meantime, this position must be held. Expect to be probed throughout the night. Keep your men alert and good luck to you."

With that Lieutenant Cross began the hazardous low crawl and mad dash back to the inner perimeter of the Con Thien firebase. Teddy, along with the remaining three men in his squad locked and loaded, waiting for the enemy to come.

Con Thien Firebase was a former Special Forces compound three miles south of the Demilitarized Zone. It overlooked a series of clandestine supply routes the North Vietnamese forces used to arm the guerillas fighting in the south.

The locals called Con Thien the "Hill Of Angels" The Marines called it "The Meat Grinder." Shortly after Teddy's unit arrived the North Vietnamese shelled the base with over three hundred rounds of 81mm mortar and artillery fire. Sappers, armed with Bangalore torpedoes and satchel charges probed the perimeter on a nightly basis. Marine casualties were high, several Medals of Honor were awarded and Teddy became a squad leader by attrition.

"We hangin' out here pretty far, Teddy," Malcom Dobbs, a private in Teddy's squad said. Their forward firing position set twenty-five yards beyond the camp perimeter.

"Tip of the spear, Mal. Better scope of the countryside from here," Teddy explained. "We see it, shoot it, and reload. Orders are orders."

"Still pretty damn far out," Mal muttered.

Teddy had three men left in his original squad. He had been promoted to squad leader when Corporal Barnes was killed. The squad had been whittled down to four after days of bitter fighting around the firebase. Now all patrolling had been suspended. The Marines had withdrawn into their perimeter and a large attack was imminent.

There were still two or three hours of daylight left so Teddy settled himself in against the sandbagged wall and drew out the now filthy and stained letter he had received from Chris a week earlier. Checking once more to see that Dobbs and

Richards were guns up and alert, he unfolded the letter and read…

Hey Teddy,

Word has finally reached me about your decision to join the Corps. At first I was pissed, but considering my own decision I guess I understand. Take care of yourself little brother, I know more about how the Corps works and I want you to listen to me. Besides, I outrank you now, I made sergeant a couple of days ago. Remember rule one in the Corps, always listen to your sergeants. I've heard from Connie and Dad and they are worried about you too. Write home as often as you can. Consider that an order. Ted, I know we didn't have a lot of time to talk when we were growing up. I was angry most of the time and away from the house as much as I could. That was wrong. I should have been around more. Now we are both Marines. I'm proud of you for that but worried as well. I've done my time over there and I know what it takes to survive. Trust your buddies. Only have buddies you can trust. Hang out with the smart guys, the ones with clean weapons and alert minds. Don't get drunk. Stay away from drugs, that includes Dexedrine, just because they give it to you doesn't mean it's good for you. Don't volunteer, there are enough ways of getting killed over there without asking for it. Change your socks, keep your feet dry, your head down and your mind sharp. I wish

I could tell you more but you'll have to figure a lot of things out for yourself.

I'm on a light duty medical profile at Camp Pendleton. Write when you can. With any luck we'll see each other in ten months, I'll make sure of it. I love you little brother and I should have told you that more often.

Take care, very good care, Chris

Teddy carefully refolded the letter and tucked it under his flak vest. Ten months. Teddy was more worried about the next ten minutes.

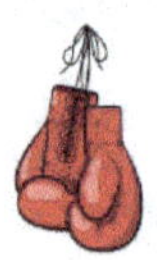

Max spotted the nattily uniformed sailor crossing the gym floor from behind the heavy bag he was steadying for a fighter.

"Hold it a minute," he told the fighter, "I think I see a ghost."

Butchie, now Kevin, now Seaman First Class Martin, sauntered into the gym carrying his canvas gym bag and a paper sack, rolled closed at the top.

"Hi'ya Max, You still punchin'?"

"Always punchin' kid, it's good to see you." Max and Butchie wrapped their arms around each other and danced away to get a good look at one another. Butchie was still in fighting trim and was dressed in his immaculate, blue and white, shore leave Navy uniform. On his chest he proudly wore his silver dolphins, an insignia that indicated his qualification as a submariner, a highly qualified and respected member of a submarine crew.

"You swimmin' underwater these days?" Max asked, noting the submariner badge.

"USS Pargo, Nuclear Attack Submarine SSN650," Butchie replied proudly.

"Now what possible use could a Nuclear Attack submarine have for a pretty good middleweight?" Max asked.

"What do you mean "pretty good middleweight?" Butchie laughed.

"Seriously kid, you got a good bunk?"

"Sonar Tech, working on my petty officer rating after my next cruise," Butchie answered with distinct pride.

"Goin' anywhere colorful?"

"Under the ice, Max. It's a big deal, Artic Ocean, could be a couple of months."

"That'll make ya', kid. I know, I knocked around with those guys for a while myself."

"I know you did, Max. Brought you this, thought you might want to wear it."

Butchie unrolled the paper bag and produced an official seaman's cold weather issue sweater, brand spanking new and Navy Blue. Max's eyes watered when he saw it. Butchie presented it to Max as a gift of gratitude and respect. Max received it as a gift of friendship and love. The old man shuffled his feet and turned his head to wipe a tear from his eye.

"Whaddya' wanna make an old sailor act like a little baby for?" Max croaked.

"I thought it would come in handy until you got a little heat in here."

"Gather round here you guys." Max signaled to several fighters who had been watching the exchange between the old friends from the corners of their eyes. They shuffled over, standing in a circle.

"This here is Butchie Martin, he shoulda' went all the way to the Nationals in the Golden Gloves but he got robbed on a split decision in Connecticut," Max announced to the impressed fighters.

"That guy who robbed me had a left hand like a Missouri mule," Butchie replied.

Several of the younger fighters knew Butchie by reputation as one of Lowell's best boxers. They stood around bashfully, until Butchie announced, "How about I get a work out, Max. I feel the need to slap leather."

"You slap anything around here you think won't slap back", Max laughed and Butchie headed for the locker room.

The rhythm of the gym resumed, the same only different, like the times.

In the quiet of the seminary chapel Peter Rayburn bowed his head in prayer. Now in the second year of preparation for his vocation, Peter found peace within himself. He had reconciled his doubts with faith and replaced his skepticism with a pious belief in the doctrines of the Catholic church.

As the outside world raged with political controversies and anti-war protests, Peter sought sanctuary in the teachings of Catholicism. He now strove to empathize with the conflicting judgments of society rather than judge them, and to pray for reconciliation of these views rather than the triumph of one side over the other. He prayed as well for absent friends, wondering where life landed Dave, Margaret Mary and Teddy. Their paths as well as his own, totally unforeseen in the dreamy aspirations of adolescence. He recalled with fondness the days when tunes by the Beatles were of critical importance, when wins or losses by the Red Sox were occasions for happiness or gloom, and when messages from the TV or the movies provided all the necessary direction for their lives.

Peter, now Brother Rayburn, closed his prayer missal and approached the chapel altar. There he lit three candles, one for each of his friends and for the faith, hope and charity he hoped they would find to guide them on their journey.

"

To improve is to change; to be perfect is to change often.

Winston Churchill

Chapter Five

Hello Vietnam

After four days and three nights our C-130 cargo plane rolled to a halt on a dark and scary airfield somewhere in South Vietnam. It was sometime past midnight, November 14th, 1967, we guessed, having endured multiple time zone changes, International Date Lines and thousands of miles of Pacific Ocean islands, stop overs and refueling points on our flight from Andrews Air Force Base in Maryland to the Republic of South Vietnam. We had not been told of our exact final destination, Top Secret and all that, but we were here, wherever here was.

The 571st Medial Detachment had deployed, carrying twelve officers, thirty enlisted men, six large metal shipping containers and forty or so duffel bags to our duty assignment in South Vietnam. No stewardesses or plastic cup cocktails on this flight. Strap seats along the fuselage and sleeping bags atop the cargo containers along with cardboard boxes with sandwiches and an apple were the amenities on our flight.

Once we were parked on the runway the roar of the plane's engines stopped, except for inside our heads. Slowly screeching hydraulics lowered a ramp at the rear of the plane and the pungent, humid smell of Asia rushed in. First there was the odor of aviation gas and hydraulic oil, then a stench we would come to know as jungle rot. As we stepped into the night the heat enveloped us in a damp, clammy surge, a cloying tropical wetness that cloaked us like an unwanted blanket. That would pretty much be the weather report for the next 365 days.

"Dave," Woody asked in his full West Virginia twang, "Are you scared?"

I looked at Woody, whose eyes were as big as saucers and wondered if I looked the same.

"Yeah, a little," I answered. "Stay close, okay?" Woody nodded, we grabbed our gear and stood up. Peeking out of the gaping hydraulic hole like startled owls we poked our heads outside. Major Cronin, along with Harry Giles, who had flown ahead advanced party for reasons too amusing to go into here, waited to greet us. They arrived a week earlier to prepare our way and get to know the ground we would occupy.

"Yo, Dave, yo!" Harry gurgled with a huge grin.

Major Cronin greeted us as well, a bit more formally, but with a smile as well. "Good evening gentlemen, welcome to South Vietnam."

"Let's form up here, two ranks," Sergeant Johnson demanded as we milled about outside the plane, pretending he was in charge. We assembled in ranks and Major Cronin addressed us…

"You are in Nha Trang. This is a coastal city approximately 275 miles north of Saigon. Nha Trang is the headquarters for First Army Field Force and home to the Fifth Special Forces. It is large, well manned and well protected. We have been here a week and have had no attacks or unpleasantries other than Private Giles's legal paperwork."

This remark got a grin from all of us as Harry's semi-legal, overly idiotic litigation status necessitated his hasty departure to Vietnam. Major Cronin generously allowed Harry to accompany him as advance party to Vietnam to forestall his looming interpersonal woes.

"The local time is 1:53 AM. Trucks will take us to our barracks shortly. There are showers available. Get a good night's rest and in the morning, we will start our mission. Formation will be at 0800. Sergeant Johnson, take charge of the men, officers to me. That is all."

So, we clambered aboard waiting deuce and half's and travelled a short distance to a barracks, a real barracks. It had electric lights, bunk beds, a shower room and latrine, just like home, sort of. Exhausted, we rolled out mattresses, shucked our boots and collapsed onto our bunks. Our in-country tour had begun.

The Vietnam combat tour was 365 days. Survive the countdown and you go home. Our clock started when we lifted off at Andrews Air Force base. Four days flight time, 361 to go. The next morning, we were rattled from our bunks at 6 AM by a platoon of Green Beret's doing their morning five mile roadwork while singing at the top of their lungs. I don't recall the tune but there were snatches of blood, guts and glory that brought us out of sleep and into awareness of where we were.

"Yo Dave, let's go get some breakfast." Harry was up, fully dressed and bouncing around my bunk like an unrestrained puppy. This bunk was the first time I had been fully horizontal in four days. Even without a sheet, blanket or pillow it felt marvelous, except for the Harry bouncing around part.

I slowly raised my head and took in my surroundings. Basic issue Army barracks, two rows of five bunk beds, green tile floor, center aisle with red cigarette butt cans nailed to posts, bare light bulbs dangling from the ceiling and half-awake soldiers in various stages of dress, undress and confusion moving about. It could have been South Carolina, Georgia or Fort Dix, New Jersey except for the heat, the smell and the haunting sense of "anywhere, any place, any time", beware.

"Yo Dave, c'mon, best we get there early, before the big line."

I assumed he meant the mess hall, if there was a mess hall, I knew all about early.

"Get Woody," I croaked as I pulled on my boots. Like most of the men I had slept fully clothed in brand new jungle fatigues. Hunger overruled hygiene, at least for now, and within minutes Harry, Woody and I were off to the mess hall.

When we stepped out of the barracks, I discovered we were on the extensive grounds of the 8th Field Hospital, adjacent to the Lon Van air base. The hospital was supported by the 254th Dustoff, a unit identical to ours with whom we would in-country train. All this information poured out of Harry as we ate a more than decent breakfast of powdery but puffy eggs, sausage, toast and lots of coffee. In a week Harry knew most, if not all, of the ins and outs, do's and don'ts and who's who of the general area. It never took Harry long to discover the lay of the land, make friends, and discover where all the flaws in the system were. And to ultimately become one of them.

The folks we ate with in the 8th Field Hospital Mess Hall greeted us with curiosity and pity. The dining area was soon packed with hospital patients, most shuffling about in light blue pajamas and white cloth slippers. And bandages, lots of bandages.

Everything looked clean, antiseptic and near normal except for the look in the eyes of the patients in the blue pajamas.

And so it went for our first couple of days in Nha Trang. We melded with the 254th Dustoff and traded off several of our personnel for veterans from their unit.

We began flying med-evacs, first as observers as more experienced crewman handled the in-flight tasks. Gradually our crew chiefs and medics began flying missions with our pilots, learning the ropes, handling the tasks, doing our job.

The Dustoff mission called for a four-man crew, Aircraft Commander, Pilot, Crew Chief and Medic. We very quickly learned that the red and white crosses indicating we were unarmed air ambulances meant nothing to the enemy and our ships were routinely prime targets of enemy gunners. After several such incidents Major Cronin called for volunteers from the unit to fly with the crews as patient protectors, giving, when possible, cover fire for the ship and protection to the wounded.

I stepped forward along with my buddies Woody and Johnny Kane, neither of whom would ever have been required to fly missions. Woody's job was to fix radios, Johnnie's to keep the unit's vehicles running, I was administration. We would be required to maintain these responsibilities and fly when needed. All six patient protector positions were filled on the first day, by volunteers. I have always been enormously proud of the men I served with but never more than on that day when six of us decided to go beyond what was expected and do more.

As we settled in there was yet another very strange piece of business to complete in those first days in Nha Trang.

"Pass these out, one each to the men. Tell them to fill them out and turn them back to the orderly room as soon as they are done."

Sergeant Johnson handed me a stack of mimeographed forms. They were headlined at the top of the page,

Last Will and Testament of

____

The rest of the page was blank except for a line that said we could continue to write on the other side if we needed to.

I passed out the forms and sat alone with mine for a while. The looming possibility that we could be killed at any time had already been driven home to us. Make no mistake, this was Vietnam, and Vietnam was an everywhere deadly place.

Only two days earlier an incident occurred right here on the base, where we harbored the idea that we were safe, that more than proved the point.

Woody, Johnnie Kane, myself and a few other guys were on the flight line working on our ships. It was around ten in the morning; the heat of the day had not turned the airfield into a frying pan just yet. We were joshing and joking when a chopper delivering medical supplies landed on the hot pad. Four or five hospital orderlies rushed out to offload the ship. One guy was carrying a chest of plasma away from the ship. He had braced his back for the load and was walking straight up and stiff out from under the blades. A sudden gust of wind or trick of fate caused the helicopter's whirling blades to dip. They caught the orderly at the edge of the blade's halo, took his head clean off.

A fountain of blood shot straight up in the air from his neck hole. He did not immediately fall down, he kind of tottered there for a moment and then collapsed at the knees and fell forward onto the tarmac. His head rolled a few feet and stopped.

Those of us who were watching sat stunned. There are no words for an event like that. Nobody moved, the orderly's torso kept bleeding, then drained dry. It was like time stopped as well. The helicopter skittered off the hot pad and shut down. The pilots and crew rushed toward the body then froze. Nobody seemed to know what to do next. Eventually order was restored. The soldier's body, along with his head, was taken into the hospital. Buckets of water were carried to the pad to rinse away the large puddle of blood. The chopper lifted off to refuel. We went back to our work, speechless.

Anywhere, Anyplace, Anytime.

That was the first, the one I remember most. The one I thought of as I held the piece of paper in my hand that said, "Last Will and Testament of David O. Ferrier."

And this is what I wrote…

Dear Mom and Dad and Bob and John,

I want my soldier's insurance and any money in my pay book to be divided equally among you. I would like each of you to buy yourselves a very nice present with the money so that you will remember me and be happy.

In my closet there is a box of stuff I would like you to give to Teddy. It's just my boxing gear and baseball glove and some other junk but please see that he gets it, okay? Also, if you ever hear from Margaret Mary again, please tell her I miss her very much and hope she is happy in her new life.

If I am killed, please remember I volunteered to be here. I am among friends, and I believe we are here for a good reason. I believe I will not have lost my life but that I have given it for my country.

I guess that's all except I would like to be buried in a veteran's cemetery among other soldiers because that is what I am trying to be. I love all of you and wish I had told you that more before now.

Your Proud and Grateful Son, DF

Then I turned my Last Will and Testament in along with those of the other guys as we pretended we were ready for whatever the future held.

Chapter Six

Meeting The Natives

"Yo, Dave, this is my friend, Louis."

Harry had taken me off the beaten path, down by the hospital motor pool among the tool sheds and broken-down motor vehicles, the military junk yard if you will. A lone hooch, once a supply room of some kind had been converted into a solitary living quarters for Harry's new friend, Private First-Class Louis Martine.

I heard a lot about this guy. He was on his third tour. He spoke fluent Vietnamese. He used to be a sergeant, twice. He ran the "Gooks" the Vietnamese nationals who worked on the base, over seven hundred of them. Nobody messed with Louis, nobody was exactly sure why not.

Entering his hooch was like walking into a circus tent. A parachute canopy dangled from the ceiling like a large white cloud, there were lots of candles, clay elephants, a buddha or three and straw mats. Books filled rickety shelves and were stacked around the room. When we entered incense was burning and Louis was sitting cross legged on the floor.

Louis Martine looked more like a pirate than a soldier. He was thin, thinner than the proverbial rail with a drooping mustache that curved far below the corners of his mouth. His jet black hair was dangerously long for a Vietnam trooper, his eyes were gray stones. He wore faded jungle fatigues adorned with an assortment of native necklaces and trinkets and a buffalo hair bracelet on each wrist.

"Harry tells me you are a man of letters," He said.

"No, I didn't," Harry protested, "I told him you read a lot."

Before I could respond Louis asked, rapid fire, "Who do you read? Why him? Or her? What is the last book you've read? When? Would you like a beer? I have stronger spirits as well. Have a seat, stay awhile."

Harry plopped down with a grin and took a bottle of beer out of a cooler that had ice in it. Ice? I continued to scan the room, reading the book titles I could make out in the haze.

"In Search of Lost Time", "Sentimental Education", "Age Of Innocence", "I, Robot", "The Art of War", "Les Miserables", "Riders of the Purple Sage", and many, many more.

"You favor the French?" I asked seating myself next to Harry.

"They were here first," he replied, "If you want to understand the now, one should examine the before. But you have not answered me, Who do you read? Why?"

"I favor Mark Twain, I believe American literature starts with him, Scott Fitzgerald was a poet who wrote immortal sentences, Hemingway was notable when not beating his chest and growling. Can I have a beer?"

Harry passed me a frosty brown bottle of "333" beer, a Vietnamese brand I had never tried before. The rumor was that there was formaldehyde in it and God knows what else. Louis noticed my hesitation.

"You must not believe everything you hear about this place. You will limit yourself."

I tried the beer. Tasty.

"Your reading is provincial, are you a nationalist then?" He wanted to know.

"You read Zane Grey and you're asking me if I'm a nationalist?"

"Touche!" He laughed, delighted. "Perhaps we shall become friends after all Specialist Fournier."

What the hell was it about my last name?

"Ferrier," I corrected, "as in Terrier. That's pretty good beer."

I had another and another after that as the evening unwound. I learned that Louis was originally fast tracked to becoming an officer. He had enlisted on a four-year hitch with ASA, that's the Army Security Agency, very hush-hush, very elite, after graduating from Ohio State with a degree in Rhetoric and Philosophy. He attended Vietnamese language school in Monterey, California, graduating at the top of his class. Such was his first promotion to sergeant, which lasted only four months.

He washed out of Officer's Candidate School at that time. He said he wouldn't buy the bullshit, whatever that meant, and got shipped to Vietnam as a consequence.

"I was amazed," he recalled, "that of the thousands of troops I was surrounded by, not one of them spoke a word of Vietnamese. They were completely ignorant of any local customs, or traditions, religions, history or national politics. I could not imagine how any country could send its soldiers off to war so unprepared."

"You know what they told us about the Vietnamese in boot camp?" I muttered as the "333" beer began to kick in. "They told us "Gooks is gooks. Don't trust Gooks. That's about it.""

Harry nodded in full agreement, which was sad.

"Gooks is an ugly word," Louis replied, "Like Nigger or Kike, Spic and what have you. It's a shame the military allows it to be used so."

"What do you call Gooks?" Harry asked between swallows.

"They have names, Harry, just like you and I. I call them by their names. As a group I called them "VN's" for convenience, but even that, I allow, is discriminatory."

"Like not being able to ride on buses and stuff," Harry agreed.

"So, Ferrier, as in Terrier, is that your real name?" Louis asked.

"Why wouldn't it be?" I answered.

"Martine, for example is not mine. I added the "e" when I enlisted, also the Louis." He was sipping something green, smelled like licorice or paregoric. His eyelids were drooping.

"So, your real name is Martin and you made it Martine?" I wondered why.

"No, my real name is Martinoj. My father changed that to Martin. I added Louis to Lou, which is the name my father gave me after we came from Albania where my first name was Luan."

I followed about half of that and still wondered why.

"Try growing up in rural Ohio with the name Lujan Matinoj if you wonder why," He answered before I asked.

This was kind of a lot to take in. Louis was apparently a character of some depth. I asked the most common Nam question, "So how much longer do you have to go?"

"Personally," he responded, "I would not be going at all. I applied for permission to stay here. The Army, in its wisdom however has decided I must go in February."

"You don't want to leave?" Harry asked incredulously.

"From where I came, I have already gone," Louis answered cryptically in what sounded like a somewhat rehearsed response.

"You want to stay in Vietnam?" I asked.

"No, this country, and its people are doomed. Darkness will prevail here. When I leave, I shall go to Thailand, there to follow an Asian path."

"Why Thailand?" I asked.

"Thailand is a monarchy, and the king there loves his people. The people love their king. In Thailand I see virtue. You should try to go there sometime; it will broaden you."

"So you're gonna go live there?" Harry asked.

"If fate allows," Louis said as he began fiddling with a tall glass pipe and a velvet sack. "You have come here, I presume, to Chase The Dragon?"

Harry's head began bobbing up and down like a floatee in a fishing tournament.

"Uh, not so much," I added cautiously, "I have to tell you Louis, I've never even smoked pot before."

Louis hesitated with the small white powdery balls and bubbling water and regarded me carefully. "Perhaps then something less harsh. I have some very nicely cured Thai stick here, pleasant, not as…powerful." He turned and opened a jar withdrawing what resembled a handmade cigar.

"Better suited for your first encounter."

He paused, holding the Thai stick like a wand. I slowly nodded my head. Harry was still bobbing away like he had lost control of his neck.

Bending to a candle Louis puffed the wand alight. Inhaling deeply he passed the prize to Harry who looked like he was about to swallow the whole thing. Harry finally broke contact and passed me the smoke.

I took a tentative puff, held it as I had seen them do, exhaled and the worm turned. A fuzzy fog filled my head, I started to float, didn't and started to float again. Louis and Harry were watching me closely. I felt a silly grin cover my face. Harry passed me the joint. Wash, rinse repeat.

For the first time in longer than I could remember that little man, that voice inside my head that told me what I should be doing, what I didn't do well enough, what I should and shouldn't know, shut up. It just got blissfully quiet inside my brain. For the first time since the ramp came down on the Nha Trang runway I didn't feel scared. I didn't feel anxious, homesick, or confused. I just kind of felt good. I reached for the doobie again. The good feeling got better and better.

Soon we were all three laughing at a story Harry was telling about something I didn't really understand and positively can't remember, but it was hilarious all the same. I found that the "333" beer made a wonderful chaser after the slightly harsh smoke, formaldehyde be damned. Louis reached back into the jar, several times.

Then for some reason Louis coughed, and shook himself, and the mood changed. Somber, far from sober, he began, "I feel compelled to speak of something to you both you have perhaps not considered in your brief time here."

Harry looked confused, I was still grinning and enjoying the inner silence, Louis was grim. "This place, my new friends, is poisoned," He announced.

Harry spit out his beer, looking terrified at the bottle.

"Not the beer," Louis said patiently, "this place, this country, us, them, all of us, poisoned. You will become comfortable here, strangely complacent, as I have. Yet at any moment death could be at your door, that door, my door, any door you enter or walk out of. A mortar shell could be dropping on us at this very moment, seconds away from killing us, all three. Today I saw a krait snake under a shattered jeep, those tiny reptiles can kill you in three seconds. Yesterday they were shooting rabid rats, lots of them, all around the garbage dump."

Louis began rocking back and forth in his lotus position. The comfortable cloud I was resting on grew darker. Harry stopped bobbing. Louis continued, "If you listen carefully, you will hear a hum. Hear it?"

We did.

"Those are the generators which run the morgue's refrigerated CONEXS which store the dead bodies of our fellow soldiers until they are shipped home. There are eight of them right next door. Each holds fourteen bodies. They are usually full."

Louis refilled his glass with the mysterious green liquid. Harry held a smoldering Thai stick as if he was no longer aware it was there. My inner silence was being drowned by an inner alarm.

"How many wounded soldiers did you not look at today?" Louis asked. "The blue pajamas make them quite invisible do they not? The guards in the gun towers tell me that they are sniped at four or five times every shift and have no idea where to return fire. I have 694 Vietnamese people working for me. At least half are passing information along to the Viet Cong. They pace off buildings, draw maps of their work areas, steal whatever they dare. Many of them do not want to do this but if they do not the Viet Cong will kill their families, then them."

"There are no good guys here. We are dropping thousands of bombs on terrified villagers, defoliating countless acres of jungle with defoliants we know nothing of the long-term effects of. Do you have any idea of what indescribable acts of cruelty a frightened, homesick, angry teenager can commit when given permission to do so? Search and Destroy, Rolling Thunder, Body Count, Snake and Nape, Arc Light, the list is endless, the obscenity profound."

Louis stopped talking, started sobbing. I had no voice. Harry was still as a statue. All we could hear was the hum of the morgue generators.

"Poisoned, my companions, death is all around us in forms you cannot imagine, in places you must frequent over and over again."

Louis shook himself, straightened up. "Apologies, my friends, I am a poor host and do go on so. One last stick of Thai to set you on your way and lighten the mood. I fear I have entered the Valley of the Shadow once again."

Louis dipped again into the jar, fired up the stick and passed it around. I drew deeply, trying to regain the solace I had felt earlier. I gained only melancholy, lots of melancholy.

We made our goodbyes and Harry, and I stumbled back to our barracks. I barely made it to my bunk, collapsing in a heap and a haze. Woody came over to talk but I was so far gone in the smoke I couldn't reply. My eyes closed. My mind floated. The little man inside fought to take charge once again. I fell into the darkness and passed out.

Two hours later the first incoming round tore through the roof of the barracks next to ours killing three soldiers as they slept. In rapid succession four more rounds pounded our unit area. More men were injured scrambling from the barracks, one of which was aflame. I had not awakened.

"Dave! Dave!" Woody was shaking me frantically. "C'mon Dave, we got to get out of here!" I flopped like a broken doll, high, stoned, wasted, worthless.

Woody tried to grab me under the arms and lift me from my bunk. More explosions rocked the area, shrapnel tore through the building. Woody threw himself across me and waited for the banging to stop. Eventually the impacts shifted to the airfield. Woody dragged me to the door. My legs, like my brain, wouldn't work. Woody half carried me to a bunker and set me down inside. People stepped over me in the darkness. Slowly I became aware of my surroundings but didn't much care what was going on.

The attack lasted for over an hour. Thirty to forty incoming rounds ripped the airfield, the barracks area and the hospital wards. Ten soldiers were killed, another twenty-four wounded.

I woke up dazed, confused, unharmed. Thanks to Woody. I also woke up ashamed. I had put my friend in harm's way. I had needlessly and carelessly endangered my life and those of my fellow soldiers. Harry, only slightly more coherent than I was told me what Woody had done. How does one say thanks to another for such a thing?

"Woody," I began, "I am embarrassed by how wasted I was last night. I can never thank you enough, but I can tell you this, it will Never Happen Again, and you will always, always have me in your corner when you need me. Thank you, Woody."

I learned another important lesson in how I must behave if I was going to complete this perilous journey I had begun.

There is one more tale to tell of our early days as a unit and of my ongoing education in the military. The story revolves around Harry Giles whose legal and personal troubles emerged as we were getting ready to depart for Vietnam.

I wish it was more a story of redemption rather than rescue but stories tell themselves and this is why Harry was at the bottom of the ramp when the rest of us arrived in Vietnam, and of how I came to recognize and accept others as they are, rather than how I wanted them to be.

The problem, the solution and the outcome are all part of the Legend of Harry Giles, and this is how the legend came to be.

THE LEGEND OF HARRY GILES

Harry Giles was, to put it mildly, a friendly, fun loving character, verging on screwball, but in a good way, mostly. He was cheerful, impulsive, faithful, and based on his life choices, not too bright. Harry was eighteen when he got to Vietnam, like most of us, but those eighteen years didn't smarten him up very much. Nevertheless, he was my friend and as we would come to learn, a very good medic. He had a way of not just treating the wounded but of comforting them. His quick grin and cheerfulness, combined with what must have been a generic medical talent, put many a terrified casualty at ease and properly medicated on his flight to the hospital.

When Harry was in the helicopter, he was one of the best, ten feet away from the ship he was generally a disaster. I know that sounds like an odd combination but that is how there got to be a "legend" around Harry Giles.

The story behind the legend began on a very busy, stress-filled day at Fort Meade, Maryland. As we were getting ready to deploy to Vietnam it seemed a thousand things needed to be done, right away, if not sooner. Major Cronin worked twelve, fourteen hours a day, most of the unit did as well, right alongside him. As soon as one problem got solved, two more seemed to arrive.

I was typing as fast as my two index fingers would let me. Major Cronin was on the phones, on the flight line, in the supply room, at base headquarters, seemingly all at once. Life was hectic, but life was good. Then one morning two grim-looking staff sergeants I had never seen before entered the orderly room. They said they were from the Army's Criminal Investigative Division. I asked to see some identification. They had it.

"Do you have a Private Harry Giles assigned to this unit?" One asked as if he already knew the answer.

"I nodded, but before I could speak the other said, "We need to see him immediately." Not a request, an order.

"May I ask what this is all about?" I said as respectfully as possible.

"You may not," one growled, "Just get him over here."

Which brought Major Cronin out of his office.

"I am the commanding officer here. If you need to speak to one of my men I am the one you should be talking to, not my clerk." He did not sound happy. The two sergeants stiffened, then in a more respectful tone the other one said,

"Our apologies, sir. We did not know you were in your office."

"And you did not make much of an effort to find out," Major Cronin replied. "Now, what is this about?"

"We have orders to place Private Giles under arrest."

The other one-handed Major Cronin some paperwork which he scanned quickly. "Private Giles is not here at this time. I will review this matter as time allows and make him available then. That is all gentlemen."

Major Cronin turned and went back into his office. The two sergeants stared at me, then left. There was silence in Major Cronin's office for several minutes, then he called me in.

"Where is Giles?" he asked.

"Base hospital sir. He's on detail, giving inoculations I think."

"Do you know anything about this?" Major Cronin pushed the papers across the desk. It took me a couple minutes to read them. I understood the message if not the legalese.

The gist of what I read, loosely translated, was this: PFC Harry Giles was being charged with bigamy. The papers asserted that Harry, in addition to having a wife, Linda and daughter, Cindy, in Philly, had married a certain Belinda Ramirez at Fort Sam Houston, Texas while in advanced medical training. Linda, Belinda? What are the odds?

I shook my head in wonder and handed the paperwork back to Major Cronin.

"First I've heard, sir."

"Go get Giles. Bring him here." Major Cronin sighed an all too familiar sigh. I high tailed it out of there.

"Yo, Dave, yo!" Harry greeted me cheerfully when I located him at the base hospital. He was giving inoculations to a bashful looking line of soldiers.

"Clap shots," Harry explained when I separated him from the detail. Harry was his usual bouncy and buoyant self as we walked out to the jeep.

"Yo Dave, we goin' to the airport? Pick up some cargo or something?" Airport runs were Harry's favorite errand with plenty of time for sightseeing and goofing off.

"No airport, Harry. Major Cronin wants to see you right away."

"Am I in trouble? Yo, Dave, what?" Harry's normal good mood was fading fast, replaced with panic and a guilty conscience.

As we drove along in the jeep I asked, "You're still married to Linda, in Philly, right?"

Harry started nodding frantically. "Is she alright? This isn't about me taking her car, right? I mean, if we're married it's like my car too, right?"

"She's fine, as far as I know and this has nothing to do with a car, any car. Who's Belinda?"

Harry appeared momentarily startled, then answered, "Yo, Dave, she's like my other wife. She lives in Texas. Is she alright?"

I pulled the jeep over to the side of the road.

"Other wife?"

Harry patiently explained, "I got married to Belinda when I was at Fort Sam. She's like a telephone operator or something."

Harry was explaining this as if there was nothing out of the ordinary here. Texas wife, Philly wife.

"Harry, apparently your Texas wife just flew up to Philly to meet your parents and tell them how happy she is that she is having your baby. Any guesses as to who answered the door when she arrived at your parent's house?"

Harry thought this over for a mini-second then turned white, not Caucasian white, Casper the Friendly Ghost white. I thought he was going to jump out of the jeep. He sat there stunned and silent, with whatever he used for a brain working overtime.

"Yo, Dave, Belinda's in Philly?" Harry's voice was about two octaves higher than usual.

"I don't know where she is now, Harry, but she was in Philly all right. She met Linda, they are both divorcing you and the Army wants to put you in jail for bigamy."

"Bigamy? What's bigamy? I didn't do anything wrong!" Harry sounded as shaky as his legal prospects.

"Harry," I asked, "Why would you marry another girl when you already had a wife back home?"

Harry squirmed around in his seat, flushed, embarrassed. "Yo, Dave, she said she wouldn't give me any pussy unless I married her." This was, for Harry, a sincere plea for understanding.

"Didn't you know, even think, that being married to two women at the same time was against the law?"

"No man, yo, we got married in Mexico. If you get married in Mexico, it's only legal in Mexico."

I couldn't think of a reply. Harry was certain he was right. He went on to explain that the other guys in his platoon told him so. He further explained that that was how all those Hollywood movie stars did it. You get married in other countries. I started to ask another question. I started the jeep instead.

"When we get to the orderly room just tell Major Cronin you didn't know it was against the law to get married twice. Maybe he will have some mercy on you."

The rest of the ride was silent and grim, which was the high point of the events to follow that day. When Harry came out of Major Cronin's office, he still had that Casper the Ghost complexion over a very worried look. Major Cronin read him the charges against him and sent him to the Judge Advocate General's office to lawyer up.

"Sergeant Johnson, would you come in here please?"

Sergeant Johnson went into Major Cronin's office. There was a long, hushed conversation. Then Sergeant Johnson came out and told me, "Prepare transfer orders for Giles. He'll be going to a holding company here to face charges."

I wondered how long you could go to jail for stupidity and started preparing the paperwork. Nobody was happy when they learned about Harry's predicament. In the short time he had been in the unit he had made friends, worked alongside the rest of us and been a positive presence in the barracks. That all seemed to be coming to an end.

And then we met Captain Horatio Marks of the Judge Advocate General's office.

"Outrageous! The boy has been charged far above his indiscretion. We shall see this right!"

So spoke Captain Marks when he came to call on Major Cronin. I sat at my desk; ears perked up as I strained to hear the inner-office conversation.

"There are two wives, two children and one husband. I am failing to see a hole in the case," Major Cronin replied, with a sigh, I imagined.

"I'll need to see Private Giles entire 201 file, financial records, medical records, copies of any awards or commendations, that sort of thing." Captain Marks asked.

"Have Private Ferrier give you anything you need. Keep me posted." Major Cronin dismissed the Captain, and he bounced over to my desk.

"How soon can you get me Giles records?" He asked, rather politely I might add.

"This afternoon. I'll have to go over to finance, the hospital, the rest are here."

"Could you bring them over to my office?" He asked. I did.

"Captain Marks," I said as I dropped Harry's records off, "Harry is my buddy. He's not the brightest guy in the world but I don't think he intended to break any laws. If I can be of any help to you with his case, please let me know."

"Do you know anything about the law?" He asked.

"I went to Suffolk University for three semesters."

"No kidding? Good school. What happened?"

"Nine ball, Mateus Rose and girls in black sweaters."

"And an irresistible urge to serve your country?" He countered.

"That too," I answered, not entirely ironically.

"Me too, grabbed my ass right out of law school. I passed the bar and got my draft notice the same day. Forever may it wave."

"How long have you been in the service?" I asked.

"Two years, four months, 12 days. How about you?"

"Six months and counting."

"Tell me about Giles, is this guy worth the effort?"

So I tried to explain Harry Giles. Happy-go-lucky, without the lucky part, friendly, dependable, (so far) and generally clueless, but likeable. Harry wasn't a crook, even Teddy wouldn't have thought so, and he had a lot of growing up to do, we all did, and Vietnam was our immediate future. We were going there together, and I felt I could trust him. He was, in my opinion, worth the effort.

"Okay," Captain Marks decided, "let's get this idiot out of trouble."

Over the next two weeks Captain Marks secured a continuance on Harry's charges while he reviewed evidence and worked with Major Cronin to shelter Harry from the particularly zealous prosecution. Harry, meanwhile, did his best to grow a halo, military wise, and mend very burning bridges with his wives, both of them.

"Yo, Dave, yo, how do I make Linda stop being mad at me? She keeps hanging up when I call her and Major Cronin won't let me go home to see her," Harry asked, still genuinely puzzled at what all the fuss was about.

"Major Cronin's not keeping you here Harry, the Army is keeping you here. You could go to jail for what you did." Sometime liking Harry took a lot of patience. "Do you understand that you committed a crime?"

"Yo, like I didn't mean to, Dave, I just wanted to, you know, get me some pussy." Harry mouthed this weak-ass excuse with a growing sense of fear at the consequences, a very slowly growing fear. "You really think they'll put me in jail?" The shake in Harry's voice was real, as was the dawning of the reality of his predicament.

"Captain Marks says the JAG wants one year in the stockade and a dishonorable discharge."

My words hit Harry like a hammer. He turned Casper white again and stammered, "Yo, Dave, I didn't do nuthin, I was just tryin' to get me some pussy."

"Stop saying that! You hurt a lot of people, Linda and Belinda, and your daughter, and the baby Belinda is carrying, and you broke the law and yes, they want to put you in jail for that."

Frustration overwhelmed friendship as I struggled not to physically shake some sense of reality into Harry. Meanwhile he looked at me like a puppy who had just peed on a forbidden floor. "Yo, Dave," he said sadly, "I fucked up huh?"

"Yes, Harry, you did."

And then we fixed it. Kind of. Three nights later I was working with Captain Marks in his office on some kind of defense or mitigation process for Harry when I noticed something in Harry's records which we had both overlooked. By this time Captain Marks and I had become friends and had dropped much of the military formality between one another. After several days of addressing him as "Captain" or "Captain Marks" he said, "Listen, Dave, when the aristocrats aren't around, you can call me "H" okay? That's what my friends call me."

"H?" I answered.

"Yeah, short for Horatio. You'd be surprised at the number of people who stumble over that name."

"I never met a Horatio before. Cool name."

"Gave it to myself in law school. All it takes is two pieces of paper and a rubber stamp at City Hall. You gotta' love the system."

"You named yourself Horatio?" I asked in wonder.

"Yeah. You know what my parents named me? Edgar. Edgar Adams. I decided I needed something with a little more flair."

"Horatio has flair," I agreed. "Didn't that piss your parents off?"

"Momentarily," he replied. "They got over it. Let me tell you something, Dave, you've got to please yourself. You wait for the world's permission or approval you've got a long wait and an uncertain outcome ahead of you."

That concept was going to take some reflection and resolve, two items I didn't have a lot of time for in my present circumstance.

"H" and I were enjoying coffee when a thought occurred to me,

"H", a guy reported into our unit yesterday and Major Cronin had to have him reassigned because he hadn't turned eighteen yet. Seems you can't send a guy to Vietnam if he's under eighteen."

"Yeah, Geneva convention and all that. You cannot assign a minor to a war zone. Seventeen is a minor, tried and true. What's that have to do with Harry?"

"His marriage certificate, the one from Juarez, look at the date." I passed him the document.

"June 17, 1967," He read.

"Here's Harry's enlistment record. Look at his birth date."

"July 4, 1950." And "H", now Captain Marks, lit up like a candle.

The grand finale of the entire episode was that Harry, on the date of his imprudent Mexican marriage was technically, a minor at seventeen. Belinda, it turned out, was a ripe old twenty-two. "H" smelled an annulment and a settlement.

"The repercussions of this whole affair are going to be primarily financial," Captain Marks explained to Major Cronin, Harry and myself. "Private Giles being a minor at the time of his marriage, could not enter into such a union without the permission of his parents. They have clearly expressed their disproval."

Everybody took a moment to glare at Harry who wisely stared down at the floor.

"Yesterday I made a settlement offer to the JAG which they have accepted. The terms are non-negotiable. Private Giles will pay Linda Giles $120 a month alimony as she is seeking a divorce. Your daughter, Cindy, will receive $80 a month child support. Belinda Ramirez will not seek alimony but her as yet unnamed child will also receive $80 a month in child support. Your marriage to Miss Sanchez will be annulled. All criminal charges against Private Giles will be dropped. He will remain civilly liable if any of the conditions of this settlement are not met. Major, how soon can you get Private Giles out of town?"

"Three days, advance party to Vietnam, with me," Major Cronin replied.

"Good, all civil actions will be held in abeyance while he is in Vietnam. Private Giles, try not to do anything foolish for the next 72 hours, then good luck to you in Vietnam."

Captain Marks, "H" to his friends, then took his leave. Major Cronin scowled at Harry and said, "Get your gear together. You will fly with me Friday to Vietnam. The unit will follow a week later. You are dismissed."

He meant both of us. We got the hell out of there. Major Cronin could have thrown Harry to the wolves. He didn't because he was a good guy. Not Harry, Major Cronin. He took care of his men, even when his men where idiots. He always kept in mind that despite our rank, our pay grade or our position most of us were kids. He would turn us into men.

Chapter Seven

Silent Night

"You ever think about maybe getting killed over here, Dave?'

Woody asked the question as we sat on a sandbagged wall along the flight line of the Nha Trang airfield. It was sunset and soon a halo of floating flares would light the night sky around us making our world an even more frightening place than it was in full daylight.

"I try not to Woody, not a lot anyway. You scared, Woody?"

"I was, when we first got here, but now I'm kinda' getting' used to it, like maybe I'm not as scared as I'm supposed to be."

I had to think about that for a minute. Here we are danger close, danger everywhere, and we're sitting outside talking like we were on a park bench at Shedd Park. Weird.

"I heard there is going to be a truce for Christmas," Woody declared, shifting a nasty wad of Skoal tobacco around in his mouth.

"That's what they say. Then at New Year's for Tet, some big Vietnamese holiday," I answered.

"Don't feel much like Christmas, does it?" Woody said as he snapped his rifle together.

"Doesn't feel much like anything except another day we didn't get shot," I answered checking the flight line, the tree line, the area around us.

"This is the first time I ain't been home for Christmas," Woody said, leaning his cleaned rifle against the sandbags. "Feels strange."

"It is strange Woody, doesn't look much like Christmas either does it?" I snapped a full clip into my weapon and leaned it beside Woody's.

Manned gun towers dotted the airfield perimeter. Triple rolls of concertina wire held us in and hundreds of claymore mines were scattered inside, outside and within the wire. Woody and I were brush-cleaning our M-16's, glancing up at the gunships swarming like bees around the airbase. It was Christmas Eve, 1967. We had been in-country five weeks. And we were no longer as scared as we should be.

"That big old Christmas tree they got in the mess hall just makes me miss being home more." Woody sighed.

"And all those Christmas cards from little kids aren't helping much either," I added.

I started to recall all those silvery Christmas mornings with twinkling Christmas trees lighting up our living room on Glenmere Street. I remembered the smile on my Dad's face as he tugged on his bright yellow baseball cap with the blue fish on it, Teddy's surprise when he unwrapped his Duane Eddy album and the burn of my cheek when Margaret Mary kissed me after putting aside her tortoise shell brush and comb set. I started to reminisce about those things, then I remembered to keep my head in the game, pay attention to what's going on around me, be careful, be alert, stay alive.

"Well at least they won't be shooting at us during the truce," Woody said wiping gun oil from his hands with a rag.

"That's something, Woody. Pretty good something when you think about it."

"Thought you said you try not to think about it? Woody grinned.

"It" took a whole lot of not thinking about. Our first five weeks in Vietnam had changed us, changed me. Our unit took our first casualty when a crew chief named Dan Lezinski took a round through his arm and another dead on his chicken plate. Woody was sitting not three feet away when he got hit. Woody said it scared the shit out of him. Scared the rest of us too, for a while. Dan was shipped home. Everybody went to chow. Life went on. We carried our weapons like they were wristwatches, no big deal, don't leave home without it. Incoming mortar and rocket rounds occasionally pounded the airfield, combat wounded filled our ships, three, sometimes four missions a day, everyday. Med-evac missions became a routine, a duty done, a job unlike any other.

The more we did what we did, the less out of the ordinary it became. The more out of the ordinary we became.

"They're havin' some kind of midnight Christmas service down at the mess hall tonight. You goin?" Woody asked.

"Nah, best not to bunch up, besides, I'm flying tomorrow. Need my rest."

"What about that there truce?"

"Locked and loaded, rested and ready. Don't believe everything they tell you Woody."

"Figure we oughta' be headin' back?" Woody asked, rising from the sandbagged wall. He stretched his arms out over his head and for all the world looked like a big bear trying to climb a pine tree. I noticed he too shot a look-around-everywhere glance at our surroundings. New habits for new circumstances, unspoken but there.

Woody no longer laughed as much as he used to, the playfulness in him that made us friends was drying up. I missed his corny jokes, his good-natured grin, his surprise bear hugs. He was still, by nature, a gentle giant, funny, smart, hard working and loyal but he had grown wary, as had I, wary and cautious as any reasonable person would be in our present situation.

"Woody, I need to ask you something." I wasn't ready to rejoin the military version of Christmas just yet. Woody sat back down and looked at me as the first parachute flares started dropping from the sky.

"If something happens to me over here," I began, shuffling my feet and not looking at Woody, "if I do get killed, would you go see my family and tell them something for me?"

This was certainly not a conversation I ever thought I'd be having, anywhere, anytime, especially not on Christmas Eve.

"I surely would, Dave," Woody answered somberly. "What is it you want me to say?"

"I want you tell them that I loved them, that I did my job as well as I could. I want them to know I died with my friends doing something good, worthwhile, you know?"

Woody thought about that for a moment, then replied, "Anything else?"

"Nah, I wrote them a letter I've got back in the barracks. I'll give to you, you know, just in case."

Woody nodded, then said, "I ain't written no letter yet, but I will. Will you keep it for me and go see my folks if something happens to me?"

"Sure Woody, I'd be honored. Shake on it?"

And we did, then grabbed our rifles and headed back to the company area hoping for a Silent Night, Holy Night when all is calm, and all is bright.

"Yo, Dave, Yo." Harry was shaking me awake from my almost sound Christmas Eve sleep. It was still full dark outside the barracks; guys were still sleeping.

"Yo, Dave, wake up. I fucked up man."

While this was not shocking news, it still took a minute or two for me to get oriented. There were no loud booming noises outside, nothing was on fire, nobody was screaming. Merry Christmas Vietnam.

"What is it, Harry?" I was awake now, sitting upright. Harry was perched on the corner of my bunk.
"Yo, Dave, I lost a jeep."

Some things are harder than others to wrap your head around when you are waking up.

"You did what?" Harry had my full attention.

"I lost a jeep, man. I went into town and parked it and when I came back it was gone."

Two things, the town of Nha Trang had been placed off limits for Christmas Eve and Christmas, also we were never supposed to take military vehicles off base without permission and never into town even when it was not off limits.

"Outside," I whispered. There was no sense in waking everybody else up with this nonsense. I pulled on my pants, my boots and my shirt, grabbed my rifle and followed Harry outside. It was always better to be prepared whenever you took a walk in the moonlight in Vietnam.

"What the hell were you doing taking a jeep into town?" I hissed, trying not to raise my voice. This was going to be serious trouble.

"Yo, man, I just wanted to get myself some pussy."

"Harry, do you have any idea how idiotic, how stupid you sound when you say something like that?" He didn't, he honestly, truly didn't.

"What am I gonna' do man? Major Cronin is going to be pissed."

Pissed was only going to be the start of it. Harry looked miserable, confused, frightened, scared, but he was my buddy.

"What time is it anyway?" Was all I could think of to say.

Harry checked his watch. "4:30."

I looked around in the darkness. "You mean middle of the night 4:30, and you want to know what to do?"

Harry didn't need to look or feel any more miserable.

"C'mon," I said, "Let's go down to the mess hall and have a cup of coffee."

The hospital mess hall was open 24/7. Coffee and sandwiches between meals, if you weren't on their asshole list they'd even let you fry some eggs if you stayed out of the way. Harry and I weren't on the asshole list but we stayed with just coffee.

"Yo, Dave, what am I gonna' do?"

"You're going to tell Major Cronin first thing in the morning,"

"What about Sergeant Johnson?"

"Fuck him. He won't even be around first thing in the morning. Your best chance is to come clean with the Major, and leave out the pussy part, Harry, just say you were drunk."

"What do you think they're gonna' do to me?"

"I don't know Harry, but it won't be pretty. I'll go in with you, help if I can."

At 0700 hours Major Cronin was behind his desk getting an early jump on the day's paperwork. I knocked on his door and he told me to come in. The first thing I noticed was that he did not look Merry Christmas happy. He did not look Merry Christmas anything at all.

"Sir," I began, Private Giles is here. He lost a jeep in town last night."

Major Cronin looked up from his paperwork. Glared up to be more accurate. "Send him in," he said curtly. "You stay as well." I no longer had to wonder what The Crack Of Doom sounded like.

Harry slunk in, drew himself up to attention. I came to attention as well, just to be safe.

"What do you have to say Private?"

Harry looked at me. I did not look at Harry.

"I fucked up sir. I lost a jeep."

"When and where did you lose a jeep, Private?"

"Last night, sir. In town."

A thundering silence hung in the room. Major Cronin shuffled some papers then held one up.

"The MP's paid a call on me last night. Seems they impounded one of our jeeps found abandoned in town last night. That wouldn't happen to be your jeep would it Private?"

"You mean I didn't lose it, Major?" Harry brightened. What an idiot.

I watched Major Cronin go from a high boil to a low simmer regarding the situation he had before him. I also learned how leadership should be demonstrated. "Wait outside Private." Harry turned on his heel and left. Major Cronin turned his attention to me.

"What do you know about this?"

"Only what Harry told me sir. He woke me up at 4:30 this morning, Told me what happened. I decided not to wake you and here I am."

"Was anyone with him when he took the jeep?"

"Couple of guys from the 254th sir. He doesn't want to name them, says this was his idea."

"None of our men?"

"Not that I know of sir. I believe him."

"Send him back in. You can wait outside."

I sent Harry in, then listened as hard as I could.

"At ease, Private Giles," Major Cronin began in a calm, level voice. "This is a court martial offense, private. You could be sent to the stockade for this."

I could almost hear Harry's knees knocking.

"Fortunately for you the vehicle was recovered, and I am going to handle this at unit level."

No stockade. I hope Harry realized what an enormous break he had just been given.

"I have had nothing but good reports on you since we arrived here. Several of the surgeons at the hospital have asked who the medic was who prepped certain patients and pointed out what an excellent job you did. Your crew chief and the pilots you fly with all concur that you do your job very well. You are saving lives, Private Giles. I need the best medics I can get for us to do that job."

And then there was silence. A full, very long minute of it. Major Cronin continued, "And then there is this. I will always, always back my men. I went to bat for you back at Fort Meade. I chose to have you stay when we traded out some of our troops with more experienced units. I had enough confidence in you to put you in one of our helicopters to administer to the wounded. And then there is this!"

"These were your promotion papers to Specialist Fourth Class, Private," I heard paper tearing. "This is it, Private. You are out of strikes, one more, one more anything and you are gone. I will see that you clean bedpans for the rest of your tour. Do you understand me?" Major Cronin's voice amped up an octave.

I couldn't hear Harry's response, but I was pretty sure even he would understand this, I hoped.

"Private Giles, you will, for the immediate future, spend every moment you are not flying, eating or sleeping right outside this office, where I can see you, is that clear? During that time, you are going to be filling sandbags, many, many sandbags." Major Cronin continued, "Then we are going to get you a stencil and a can of black paint and on those sandbags, you are going to stencil "Compliments of the 571st Medical Detachment". I am going to want to see those sandbags everywhere I go on this compound. Everywhere. Now go find a shovel and report back here. Go. "Private Ferrier, come in here please." Major Cronin's voice had returned to his usual businesslike tone.

"Permanent Private Giles is going to need some equipment," he said as he wrote out a list and handed it to me. "Take this to Sergeant Adams in supply and see he gets it."

I took the list and turned to go. Major Cronin stopped me, "One more thing Private," he handed me a set of orders, "You are now a Specialist Fourth Class, congratulations."

I read the orders. Harry's name was crossed off the list of promoted privates.

"Thank you, sir," I said, offering a salute. Major Cronin returned my salute.

"One more thing if I may sir," I said, "Merry Christmas, it's been a wild year so far."

Major Cronin smiled, wished me a Merry Christmas and returned to his paperwork.

Next year was going to get a lot wilder.

Chapter Eight

North Of Danang

The armistice for Christmas 1967 was followed by a much longer truce and cease fire agreed to by both sides, to celebrate the Vietnamese New Year of 1968. Called, "Tet" this weeklong observance was the most sacred time of the year for the Vietnamese people. Both sides agreed to stand down during the holidays. All aerial bombing halted. American and Vietnamese ground troops suspended operations and patrolling. On the first night of the arranged truce Viet Cong and North Vietnamese forces in a coordinated effort, attacked every major city in South Vietnam and every US firebase between the DMZ and Saigon. On the first day of the truce two hundred and forty-six American soldiers were killed in the treacherous attacks, the highest daily total of the entire ten year war.

By the time "Tet" was over two thousand, six hundred American and South Vietnamese soldiers had been killed in action. Another twelve thousand, seven hundred were wounded. Enemy casualties were estimated to be as high as fifty thousand.

Thousands of Vietnamese civilians were slaughtered in cities and villages overrun by enemy forces. The United States embassy in Saigon was captured, then retaken as Saigon burned. In the Imperial City of Hue over five thousand elected officials, schoolteachers, Catholic priests and nuns, Buddhist monks, university students and civilians were rounded up and executed while the city was in enemy hands.

Throughout the opening week of the attacks, we were dug in on the airfield at Nha Trang. Sappers repeatedly probed our perimeter only to be repelled by a tough as nails contingent of South Korean rangers on the front line. The warriors of the Fifth Special forces further decimated the attackers.

Our Dustoff helicopters were rarely on the ground, carrying scores of wounded to packed hospital wards. The fighting raged on twenty-four/seven for a month. Eventually the enemy was driven out of the cities, away from the firebases, back to their jungle strongholds. We packed their dead on paddy dikes.

In the midst of the carnage the 571st Dustoff received orders to pack up and move three hundred miles north to the fire base at Phu Bai. This was the staging area for the Marine Corps assault on the captured city of Hue. Both the Army's First Cavalry Division and the 101st Airborne were joining in the assault to retake the city. Casualties were expected to be high. We would be the furthest north Dustoff unit in Vietnam. We would be the only Dustoff unit north of Danang. We would be on our own.

It took five days to get our entire unit up north. Some of us flew up in our soon to be riddled with bullets helicopters, refueling at one besieged firebase after another. Others came by sea, packed into LST's with our main gear while hugging the coastline for the journey. Finally, there was a perilous convoy north from Danang through the treacherous ambush alley called the Hai Van Pass.

When we finally settled in, Phu Bai was a muddy swamp cleared for our tents, a partially constructed surgical hospital not yet ready to receive patients, and an exposed helipad the enemy made a prime target for their rockets and mortars.

Between January 30 and March 3, 1968, the Marines and Army divisions suffered over twelve hundred casualties retaking Hue. ARVN forces lost another two thousand. When the carnage stopped the American and South Vietnamese forces had won every battle, retaken every objective and some say, in the hearts and minds of the American people, lost the Vietnam war.

"Hold your hand there, press hard, change the pad when it gets soaked."

I caught a quick look at this guy when he had been loaded on board. A shock ran through me, like an icicle right through my stomach. Teddy. This guy looks just like Teddy. For a moment, then he didn't, his face twisted in pain, his eyes squeezed shut. Airborne patch on his shoulder, Army fatigues, bloody and wet, not a Marine, not Teddy. Harry scuttled over to another wounded soldier leaving me crouched next to a dying man who had a gunshot wound, upper chest, deep, arterial bleeding and miles to go before we got to the 22nd Surgical hospital.

I watched the gauze pad under my hand turn deep crimson and grabbed a fresh one from the stack Harry had tossed nearby. My fingers were sticky with blood, the wind whistling through the open helicopter doors raised a faint pink cloud around the wound, stinging my eyes, choking my throat. I tried not to look at the guy's face anymore.

There were four more wounded in the ship. Harry moved among them, checking a bandage there, a dressing there, elevating a head, checking a breathing passage, stabilizing those that he could, comforting those he could not.

"Plasma Hotel, Dustoff 508."

"508, Plasma Hotel, go."

"We are inbound your poz, 4US litter, 2 ambulatory. ETA one five."

"508, roger. Stretchers on the pad."

I could hear Pete Johnson, the aircraft commander calling us in to the base hospital. I also heard we were still fifteen minutes out. My guy wasn't going to last that long.

"How we doin' back there Harry?" Pete turned partially in his seat to check the bay. Pilots didn't do that often; it was generally not a pretty sight.

"Guy Dave's got is pretty critical. The rest are stable."

"Good job Harry, hang in there Dave."

Pete turned around, took the controls and goosed another twenty miles an hour out of the already balls to the wall helicopter. My guy stopped bleeding, then he stopped breathing. I rocked back on my heels and suddenly felt cold all over. I stared out the door of the helicopter, watched the insanely green jungle flash past and tried to stop my brain from thinking.

I had seen guys die before, up close, same way, same conditions. It didn't help. I stopped my pressure on the wound and tried to wipe some of the blood off my hands. I felt myself choking up and forced myself to stop. I felt small and helpless.

I didn't look at the dead guy the rest of the way in. I lit smoke for one of the ambulatory wounded. I was up to two packs a day myself. He leaned over to give one of the stretcher wounded a puff. I kept trying to swallow, couldn't.

Ten minutes later we were on the hot pad at the 22nd Surge. The wounded were off-loaded and hustled into the wards. We jumped back aboard, refueled and shut down in the first up revetment. I was tying down the blade when Harry came over. "Yo, Dave, I'm sorry I put you on that guy. He wasn't going to make it and I had to check on the other guys."

"Yeah, I know Harry," I whispered, "comes with the territory."

"You guys, alright?" Pete Johnson came over, checking us, checking the ship.

Harry and I nodded, neither of us were anywhere near alright.

Since the start of the Tet battle all six of our helicopters had barely been on the ground. We were flying around the clock hauling wounded. The ships got makeshift maintenance as they could. Before the fighting was over all six of our ships would be disabled, shot up, and replaced. So would many of the crews.

Mike Peters had been shot. Bruce Goodwin as well. Both survived and med-evaced to Japan. One of our pilots, Jim Seransky, broke both of his ankle's force landing a ship in a rice paddy. Miraculously nobody from our unit had been killed, yet. Major Cronin, a man of understated but fervent religion, believed an angel watched over our unit, at least up till now.

"Hey you guys, I need a hand here." Steve Camarano, the crew chief, was cleaning up the ship. Wet bandages, old field dressings, forgotten gear littered the crew bay. Harry and I turned to help. Harry swabbed the deck as best he could, blood, mud and debris sanitized before the next mission.

"Dave, your weapon's got some blood on it," Steve mentioned as he re-hung litters. My rifle had fallen over onto the bloody pressure pads I held against the dead soldier's chest. I retrieved it, cleared the magazine and began to wipe it down. The trick was to not think about what you were doing, just do it. Action, not insight. Get ready, be ready, stay ready. Focus. My weapon clean I clicked it back into its rack by the gun hole.

"Yo, Dave, let's get some coffee." Harry was wiping his hands with rubbing alcohol. I did the same.

"Steve, we good?" I asked.

"Yeah, go ahead. I'll see you up in operations." Steve continued to groom his ship, checking commo, checking latches and clips, checking wiring, checking, checking, checking.

Up in the operations shack Pete Johnson finished the mission paperwork. Rotor hours, casualties, disposition, all had to be recorded for whatever posterity was to come, or to care.
"You guys, okay?" Pete repeated as Harry, and I came in. Before we could answer, Malcomb Perry, the radio operator interrupted, "Did you get the name of that guy who died? Ground commander wants an update on his men."

Harry, Pete and I looked at one another. We generally didn't talk much about guys who died on board the ship. Malcomb was a new guy, still learning the unwritten rules. Silence hung in the room for a long, long minute.

"Gunderson," Harry said softly, "I saw it on his name tag."

Malcomb bobbed his head and turned back to his radios.

"Let's go up to the mess hall and scrounge up a sandwich. I'm buyin'" Pete suggested. Dustoff crews ate at odd hours, catch as catch can, the cooks knew this and treated us well.

"We're on the bell, Mal," Peter told the radio operator as we headed for chow. First up crews were alerted for missions by ringing a bright red alarm bell mounted on a post beside the operations shack. The ringing usually brought all activity in the unit to a halt until the nature of the mission was established.

Pete, Harry and I wandered up to the mess hall. The first pilot stayed in operations sipping coffee and talking with Malcomb. When we were seated and eating Harry said, "Yo Dave, there wasn't nuthin' we could do for that guy today. Big hole, wrong place, there was no way to stop the bleedin.'"

"I know that, Harry. That's not what's bothering me."

"What is it then?" Pete asked as kindly as he could.

"Guy that died today," I began, "looked kinda' like somebody I knew back home."

"Gunderson," Harry whispered, "Guy's name was Gunderson."

"Wasn't him, was it?" Pete asked, once again gently. I shook my head and stared at my coffee cup. "Sure, looked like him though."

"Look at the wounds not the wounded," Harry said, "That's what they told us at Fort Sam."

"Does that work?" Pete asked.

"Nope." Harry's response was short and final. He took a large bite of his sandwich.

As he chewed Dr. Marvin Stearns, one of the hospital surgeons joined us at the table.

"Gentlemen," he began, "Mind if I join you?"

Doc Stearns was a good guy, drafted right out of medical school, on his second tour in Nam, voluntary. He cared.

"Your wounded are doing fine. Nice job Harry, they were prepped well."

Harry nodded and kept chewing.

"The guy who didn't make it," Stearns added, "I'm amazed he lived as long as he did. Somebody tried to tie off the artery with a gauze strip. That you Harry?" Harry shook his head, but he stopped chewing.

"Some of the things these ground medics try to do are amazing," Stearns sipped his coffee and went on, "You guys too, Harry, sometimes they work, others, not so much. Wouldn't have mattered with that guy though."

"Gunderson". Harry whispered again, "His name was Gunderson."

Chapter Nine

Papa San

He came to work every day in the same clothes. Tan shorts, white cotton shirt, old army boots with no socks and a battered slouch hat. His clothes were clean at the start of the day, begrimed with hard work and sweat by the end.

I do not know his name. I have no idea how old he was, old though. His face and hands and arms and legs were burned brown by the sun. Deep lines penciled his face and a white scar ran from his ear to his jaw on the right side. He was missing three fingers on his right hand, pinky, ring and index. I don't know why.

We called him Papa San when we wanted him to do something. He would smile and nod in understanding how much we never knew. He always worked silently, diligently. When he rested from his labors, he often smoked a long, droopy, flat looking, hand rolled cigarette. Sometimes we saw him pick up the discarded cigarette butts we scattered around the unit area like careless coins. These he would gather and carefully shred into a pouch, saving the inch or two of tobacco we threw away so thoughtlessly.

We called him a gook when we talked about him, as in, "Get the gook to clear away all that brush around the generator," or "Don't leave the gook alone when he's working near the ammo bunker." We didn't take much notice, or even care if he overheard us saying this. He was a gook, and gooks were gooks. The unnatural order of things.

I didn't like to think of myself as a racist or a bigot. I had seen and heard enough of that crude stupidity to see it for what it was, ignorance and fear. What I did not see was the connection between how I, we, regarded Papa-San with those same two humiliations.

I never knew how to say "Hello" to him in his native language. Or "Good Morning", "How are you?", "not even "Please" or "Thank You." We communicated in a sort of pidgin, Tonto English, half pantomime, half barking. Show and Tell, as if you were talking to a small child. He would smile and bow his head when he understood, or when we thought he did. Then he would perform the chore.

From the earliest phases of our military training, we had been told two things. One was that we were on the fast track to Vietnam and better pay very close attention to what we were being told if we wanted to stay alive. Second, we were warned about "gooks". I had never even heard the term before but came to understand that these were the people who wanted to kill us, the Viet Cong, our enemy. This enemy wore no uniform, they did not have camps or forts of entrenchments where we would find them. They lived among the civilian population, farmers, merchants and peasants by day, killers by night and by ambush. Gooks.

The harsh reality that in Vietnam you could be killed, "Anywhere, Anyplace, Anytime" was further enhanced by the fact that you could be killed by "Anybody". Men, women and children carried and concealed weapons, babies were wired to explode as booby traps, soft drinks and beer were poisoned, friends were enemies, enemies were all around, never trust any of them. If you want to stay alive. Gooks.

Papa San was paid for his labors in piasters, the local currency, at the end of each week. He would smile gratefully and never count the money he slid into a baggy pants pocket. He would softly say, "Thank you", the only English words I ever heard him speak. He worked every day the base was not closed for combat, or combat alert. And I never knew his name or addressed him as anything except "Papa San".

He came to work for us as the countryside around the Phu Bai Fire Base became more "pacified". Initially there was no sewage system on the base, all human waste had to be burned, in stinking half-barrels drenched with gasoline, every day. Dark clouds of foul, fecal smoke rose every morning around the base from the outdoor latrines. Labor for this task fell to the unlucky troops assigned to the detail. No one, as you can well imagine, wanted to burn dozens of half barrels of shit every day. But someone had to.

Clean clothes were another problem. No Laundromat here. No washer/dryer appliances. We wore our fatigues until they were rank, dunked them in boiled water, hung them out to dry and put them on again.

Every structure we lived in, worked in, or hid in had to be protected by sandbags which absorbed shell fragments and blast damage from the frequent rocket and mortar attacks we suffered. Thousands of sandbags which did not fill themselves, stack themselves or replace themselves when damaged.

Six helicopters had to be manned, repaired, maintained and protected. All day and all night, every day and every night. Support vehicles, fuel and water trucks, jeeps, generators, radios and firearms had to be kept clean and in working order. All these tasks subtracted from the time we had to do our main job, saving the lives of the wounded.

So we were allowed to hire "gooks" to perform our more menial, everyday tasks Six women were hired as laundry and housekeeping workers, two men would be day labor. They would wash our dirty clothes, do our house cleaning, fill our sandbags, police up the area and burn our waste.

We kept an eye on them as they worked, never really trusting them. Our weapons, side arms and rifles were always kept secured. We didn't leave ammo or medical supplies lying around. They were searched every morning by MPs at the base gate coming to work and searched every evening as they went home. On occasion I watched this ritual, a rough shakedown by ARVN MP's who casually stole any item they wanted from them as contraband, laughing and pushing them around as they did so. American MPs stood by and watched, forbidden to interfere with the process. Papa San would endure this humiliation silently, with dignity, every single day. In all the time I spent in Phu Bai I do not recall a single instance of one of these workers being caught stealing from us. The occasional lost or misplaced item was most often found and returned, often by these workers.

Papa-San generally worked alone, or with a variety of sidekicks who never lasted more than a month or two. Whether he was digging trenches, filling sandbags, stretching barbed wire through our foul-smelling swamp or burning our shit his expression was always the same, stoic, calm, almost serene. It was as if he had accepted this lot in life and did not rail against it. He had always been treated like a common machine and always would be, by others before us, by us now. Gooks.

To survive my tour in Vietnam I had to learn to push away my fear of the terrifying environment I lived in. Hundreds, thousands of soldiers, just like me were being killed and maimed all around me, dozens, each and every day. I saw them in our helicopters, bandaged, bleeding, terrified, dying. I saw them in the hospitals, dazed, hobbling, traumatized. Somebody did this to them, somebody. Gooks. So, I feared them as an entity without regard to age, sex or personality. Gooks was gooks. Keep it simple, stay alive.

I never knew whether Papa-San had a family, a wife, children, brothers or sisters and I didn't care. It was him I was afraid of, deep down inside where I couldn't talk about it or let it show. It was him who may have been pacing off our unit area for mortar or rocket attacks, him who may have been stealing whatever he could from us and giving it to the Viet Cong. He was the only visible manifestation of who was hunting us, killing us and the docile, hardworking, respectful façade he maintained may have only been his way of fighting the war.

Through the haze of my war days in Vietnam I vaguely recalled an evening long ago with a character who named himself Louis Martine. The wisdom he imparted through the clouds of psychedelic smoke forgotten or obscured. Such is the poison of war. Distrust, fear, rumor and myth, statistics and propaganda, all in a stew served daily.

Then there was a day when Papa-San wasn't there. Another Vietnamese male, younger, smilier, livelier, reported for work in his place. I never knew what happened to Papa-San. It never occurred to me to ask. He just wasn't there anymore.

And the war went on.

Chapter Ten

Colorado Rocky Mountain High

Colorado State Trooper Rick Girard guided his four-wheel drive jeep carefully through the rutted snow. An early spring melt had given him the chance to check on some of the hidden mountain meadows recently populated by nomadic hippie tribes who often found communing with nature more challenging than they expected.

Kendall Prescott saw the trooper's black and white jeep enter the compound. "Lock the dormitory," He commanded to one of his now minions who had endured the long winter. "I'll handle this creep."

Kendall crossed to the now parked jeep as Trooper Girard got out. "Can I help you officer?" He asked as smarmily as possible.

"Just a welfare check," Girard replied taking in the run down and shabby condition of the encampment. "Is there a Margaret Mary Sullivan here?"

"Nobody here by that name Officer," Kendall replied.

"Mind if I take a look around?" Girard said as he began taking a look around. Several of the commune residents poked their heads out of teepees, tents and shacks, wary of the uniform walking among them. One young girl scurrying to a shelter was stopped by Girard.

"Excuse me Miss, do you know a Margaret Mary Sullivan?" The girl didn't answer, obviously terrified as she glanced furtively at the locked dormitory building. Girard did not miss the gesture. The girl stared at her feet, shook her very filthy head and never looked up.

"Okay," Girard said to Kendall, "You folks take care, if anyone hears about the Sullivan girl, please let the authorities know."

Trooper Girard walked back to his jeep as Kendall smirked in satisfaction. Girard turned the jeep around and left the valley. Almost.

"Dispatch, this is Girard. I am at Tuloc Meadows off State Road 74. Request two backup units, a med wagon and the county lock-up van. Come silent, rendezvous on me at 74 and I-75."

"Roger Girard. ETA on you in 45."

One hour later Girard had the whole hippie herd camped in Tuloc Meadows rounded up.

"You sir," Girard said, "You are the one who told me there was no Margaret Mary Sullivan at this location, am I correct?"

Kendall was doing his best to fade back into the pack when Girard called him out.

"Come on out here, let's take a look around." The trooper led Kendall to the locked dormitory building. The padlock and hasp on the door were the only two things the troopers had seen that were not filthy or in disuse.

The back up officers and med teams were spread out around the compound checking the other tents and shelters. State trooper Dan Evans followed Kendall and Girard to the dorm carrying a small crowbar.

"You have the key to this building?" Girard asked Kendall.

"You have a warrant?" Kendall replied.

"Do you have a deed?" Girard answered. "Break the lock, Dan."

Dan pried off the padlock and found Margaret Mary inside, tied to a cot with a rag stuffed in her mouth. She was filthy, ragged, malnourished, but alive.

Margaret Mary was placed in an ambulance and taken to a hospital. Kendall was handcuffed and taken to jail.

Jackie Ferrier quickly separated the letter from the flurry of daily mail, smoothing it flat and setting it aside on the kitchen table. The red, white and blue bordered envelope addressed to *The Ferrier Family* and marked postage *"Free"* came from me, in Vietnam. Another one, she thought, which meant I was still alive, so far. Relief, anger and curiosity coursed through her. Anger, of course, won the day. She never managed to see anything other than a rebuke in my letters, a reminder that I didn't care how much anxiety I caused her by my thoughtless decision to extend my tour in Vietnam. She would always see this as a thing I did to hurt her, not to make a decision for myself to do the right thing, even if others thought it was wrong. My Mother reacted with anger, a genetic thread I would inherit to substitute for any emotion I felt too painful.

My Mom would, as usual, wait for my Dad to get home from work before they opened the letter together. Later that evening, after supper, the family would gather around the kitchen table and my father would read the letter aloud. This one said…

May 12, 1968
Hello Everybody,

It's Tuesday here (I think) and that means I have 150 days to go. Things have been pretty quiet around here so don't worry, I've learned how to stay out of trouble and be where it's safe. Most of the fighting now is in the central highlands way south of here.

I got the package you sent, thank you very much for the cookies and the magazines. The pictures of you guys out on the boat were really cool, I showed them to the guys and promised Woody and Harry we'd all go fishing when we get home. Mom, are you ever going to go out on the boat, even for a ride? Didn't think so.

Has anybody heard from Teddy? He must still be here in country somewhere but I don't know where. I'd love to find out and maybe I could go see him Marines aren't that hard to find. I still haven't heard from Margaret Mary. Is she alright? Maybe you could call Sean and find out. I sure would like to know. Please try and find out, OK?

I hope everyone else is doing well. Tell Bob and John I picked up some really cool souvenirs when I was in Danang last month and I'll bring them with me when I come home. That should be around Thanksgiving sometime. I don't know exactly when because sometimes they give guys a "drop" and send us home early. I don't want to jinx it by counting on it too much.

In my next package could you send some black licorice? We were talking the other night about what we couldn't get over here and black licorice made the list. I know it sounds stupid but just a few sticks, if you can.

I've got to go now, my shift on the radios. Please tell Aunt Eileen I got the holy cards and the rosary and thank her for me. Also, I got a letter from Peter Rayburn, he going to Rome (Italy) to study for being a priest. It's a big deal because they only pick a few guys to do that. Good for him Say hello to everyone, I miss you and I'll write as often as I can.

Much Love, Your Son & Brother the Soldier (for now) DF

Meanwhile, in that turbulent spring and summer of 1968 both Martin Luther King and Bobby Kennedy were assassinated. Racial riots raged in the cities, college campuses erupted in anti-war protests, President Lyndon Johnson announced he "would not seek, and will not accept" his party's nomination for another term as president. Peace and war candidates vied for the vacancy while Richard Nixon lurked in the shadows.

At the movies, Rosemary had a Baby, there was a Planet of the Apes and the year 2001 promised a Space Odyssey. Both The Beatles! and Elvis missed the top ten, twenty, thirty and forty and Gunsmoke, Bonanza and The Mod Squad competed with The Beverly Hillbillies, Bewitched and Lawrence Welk for TV viewers.

And none of this, none of this, mattered as the war raged on in Phu Bai.

Chapter Eleven

Fresh New Guys

"Three kings." Fresh new guy Don Avila laid his cards down and reached for the hefty seven dollar and fifty cent pot.

"Straight to the eight." I grinned with my winner.

"Those are both real nice hands," Woody drawled, "but I've got five diamonds, all red." And put an end to any more speculation. He raked in his winnings while Don refilled paper cups with Crown Royal whisky.

"Dan, you weren't in on that hand, what happened?" I asked.

"A pair of fours is what happened," new guy Danny Hernandez, forever known as "Danny H", answered.

"Dan, it's not whether you win or lose, it's how you play the game that matters," I ventured.

"It matters when we are playing for money. A pair of fours doesn't play very long," Danny H responded. It was hard to argue with that reasoning.

More drinks were passed around. A pleasant glow descended.

Our language was a lot saltier during our Phu Bai days and I assure you, "Fresh", as in the title of this story, was not the "F" word applied to the replacements who began arriving in our unit during the late summer of 1968. They were bright-eyed, bushy-tailed, green as grass and generally scared shitless when they got their first look at Phu Bai combat base. We did not make their arrival any easier, with our kind of a frat boy swagger and "old guy" attitude. That crap generally passed quickly though. Our job was too important not to get these guys acclimated and trained as soon as possible.

Danny Hernandez and Don Avila were among the first of the "new guys" to arrive. Don arrived with the rank of Specialist Fifth Class, the same rank Woody, Harry, John and I took a year to earn in Vietnam. Don had been promoted right out of his Crew Chief training class. Why? Because, as it turns out, Don is a good guy, smart, a college graduate from Kent State University who was married with a baby on the way when he got drafted. Don was a few years older than the rest of us, a frosty twenty four, and if not more mature, at least more temperate in his immaturity, mostly.

Danny H was a Hammond, Indiana firecracker, also married, who came to replace our supply sergeant, the infamous, "Stinky", yet another short term replacement who never lived down the nickname he garnered for the colorful skunk tattooed on his shoulder and his "I gotta' get out of this place" attitude. Stinky went south to Da Nang to finish his tour. Danny showed up in Phu Bai to start his.

Danny H and Don got involved in the unit very quickly. Don, while waiting to be assigned to a helicopter, started working in Operations, keeping flight records, mission statistics, awards and decorations procedures, picking up the mail, generally helping out wherever he could.

Danny made the Supply Room a congenial "take a break" gathering place. Amidst the piles and stacks of gear Danny brought a lively sense of humor, unfiltered opinions and a steady supply of equipment we needed, some before we even knew we needed it. Working alongside Supply Officer and crack pilot Captain Ken Harman, they kept us in the essentials along with luxury items like poncho liner blankets, Nomex flight suits and aviator sunglasses.

They climbed aboard, got to work and became valuable members of the unit. Others took a different path.

"You are so full of shit, Winn." Everybody stopped the chatter and turned. This comment came from Johnnie Kane, one of the original over in the plane guys, with us in-country nine months, and usually as quiet as he was stoic. John didn't talk much, but when he did, we listened. Some people have that knack.

Fresh New Guy Dennis Winn turned to face John, unsure of how to respond. John continued, "You think any of that stuff matters here? You think anybody cares?" John was heating up, I'd seen him do it a time or two before. It could get ugly.

Winn was pontificating about the latest anti-war, political gibberish swirling though the media and the national magazines back in the world. He had been in country a little over a week and brought with him all the latest news, gossip and irrelevant nonsense being printed, published and preached about the war from people who had never been to Vietnam. Winn was an honor graduate from some Podunk junior college which he had been pried out of by the draft. He showed up at the unit carrying his duffel bag, a guitar, with a peace symbol medallion stenciled on the case and an attitude.

Winn was a medic whose future was likely to involve a lot more bed pans than flying Dustoff.

"I think what John's saying is all that political nonsense doesn't matter here, Dennis," I added, trying to throw a little oil on the choppy waters. "You've been here what? A week? John and I, Woody, Harry, we've been here almost ten months. You know what's important to me here? John and Woody, Harry and all the rest of the guys in this unit. Just us, not Nixon, not Johnson, not Woody Woodpecker or the Beatles! Just us."

"You're sayin' we'd be in a world of shit if Nixon gets elected? We are in a world of shit now," John continued, his jets cooling, just a bit. "Every day I get up and hope I don't do something careless or foolish that gets me, or somebody else hurt or killed. I don't care who's president, or what foreign policy is, I don't care who wins the Super Bowl, the Academy Awards, the World Series, all that shit is for later. Right now get your head around where you are and what your job is."

John stormed off, testier than normal, but absolutely right. The New Guys needed to get their head in the game, and fast. Their lives and ours depended on it.

We were gathered around my footlocker and bunk in the rear of one of our brand new, plywood, not on the ground, tin roofed, electrically wired hooches.

This luxury resulted from many hours of construction, by us, around our flying, maintaining, eating, sleeping and resting schedules. The Army supplied the materials, we supplied the labor. We built ten hooches for barracks, a supply room, orderly room, operations shack, and with the leftovers, a clubhouse bar and rec-room. After months of soggy, muddy, drafty tents we were inside, as good as it gets in Nam, better than it ever was in Phu Bai.

There were already two or three dozen beer cans scattered around my floor, several Dr. Pepper's as well. The gathering in my area had become a nightly ritual, established mainly by my proximity to the rear door of our hooch, the one we would all pile out of if things became dicey. The forum generally was friendly as far from home teenagers tried to talk away some of the homesickness, fear and hardships of our day-to-day existence. Philosophy was for later. Politics as well. We discussed the important stuff.

"Carol, her name was Carol Legrom. She said she'd write to me, took my address and everything." Woody was talking about one of the Donut Dollies, Red Cross volunteers, who visited the firebases with coffee, donuts and kindness whenever they could get in, and out, without gunfire.

"She's from Wheeling, that's not far from Parsons. She said she'd call my mother when she gets back home." Woody leaned forward and spit a disgusting wad of brown tobacco juice into an empty (we hoped) Dr. Pepper can.

Donut Dollies were young women of incalculable compassion who came to Vietnam to raise our spirits and remind us of our lives before we came here. Every visit was a brief time out, a moment away from the war, welcomed by all of us.

"Which one was she, the one with the red hair?" Danny H wanted to know.

"She was the one who sounded just like Woody only her ass wasn't as big," I replied, which got a laugh from everybody, except Woody.

"I heard we were going to get eight nurses, female type, assigned to the Surge," Don mentioned, sipping his now trademark Crown Royal cocktail.

"They never sent nurses this far north before," I commented, "we must be getting civilized."

"No," Don said, "Civilization doesn't look at all like this. I remember."

"Me too," Danny H added. "You guys have been here too long."

And we laughed and more drinks were had and Dennis Winn's contemporary opinions wisely faded into the background like Fresh New Guys were supposed to do.

The next day I was flying first up with Harry and Steve Camarano. We were taking two new guys with us for on the job training. It was the only way to learn.

"Yo, Dean, you're not gonna' need that, don't take these, pack more gauze, pressure pads and ace wraps. This other stuff is just going to get in your way. That's it, got it?"

Harry was prepping Dean Taylor, the Fresh New Guy medic for his first med evac. Harry, who could get himself in some kind of trouble tied hand and foot and locked in a phone booth, was, nevertheless, an excellent medic. In a crisis he was cool, calm and collected. The rest of the time he was a monkey with a machine gun.

I was showing Larry Tighe, the other New Guy, where to sit next to me in the gun hole, what to do, and how to stay out of my way while I did it. Dean and Larry were going out as observers on our next first up mission. Learn by doing on the job training, with gunfire. There was no other way to understand the process. They were trying not to look as scared as all of us felt. They were not doing so well at it.

"Look," I explained, "We go out high, above small arms range, and stay that way until we approach the landing zone. Then the pilot is going to drop like a stone, go like a bat out of hell to where the wounded are and set us downright next to them. We jump out, get 'em loaded and get the hell out of there. We've got the best pilots in Nam flying these things, the trick is don't think about what you're doing, just do your job as quick as you can and then do it again the next time and the time after that."

"How many of our ships have been shot down since you've been here, Dave?" Larry asked.

"Enough to tell you these things can take quite a beating before they go down. Mostly we get small arms fire, does body, rotor damage, but they keep flying. One of us gets shot and we got an express ticket to the hospital. It doesn't happen often, sucks when it does. Nothing you can do about it."

"I feel remarkably not better." Larry was exhibiting a very valuable sense of humor.

"You are going to feel remarkably much worse. Don't sweat it, we all do. Just do your best, remember your training and try not to fuck anything up." Same advice I was given, what felt like many long years ago.

We didn't go out for the rest of the afternoon. Some days were like that. For Taylor and Tight the waiting was just as stressful as the going but there was nothing to be done about that. Their time would come, always did.

"Phu Bai Dustoff, Phu Bai Dustoff, Makin Taxi Victor, over."

Around sunset the mission came in. The radio operator hit the bell and we headed for the radio shack.

"Taxi Victor, Phu Bai, Go."

"Roger Phu Bai, I have an emergency med-evac, 2 U.S. Gunshot wounds, litter, One U.S. Gunshot wound ambulatory."

"Taxi Victor, Phu Bai, Wait one."

As soon as we assembled the radio operator took the field coordinates as the pilot and Aircraft Commander plotted their location on the wall map.

"Twenty minutes, Andy," Stu Willis the aircraft commander relayed to the radio operator. Harry and I headed for the aircraft followed by Taylor and Tighe. Steve was already on the pad checking the ship.

"OK, listen up," Steve said as we loaded our gear aboard, "Tighe I want you in the gun hole on my side of the ship. Don't do anything unless I tell you to. Got it?"

He did, we did, and we lifted off, skids light into the darkening sky.

"Makin Taxi Victor, Dustoff Five Zero Two."

"Zero Two, Taxi Victor Go."

"I am inbound your poz, ETA two zero minutes.

"Please advise last enemy contact."

Makin Taxi Victor told us the Landing Zone was secure and would be marked by strobes. Last enemy contact, he said was thirty minutes ago. Last contact didn't mean they were not still around, waiting.

Back in Phu Bai the radio shack was filling up. Whenever a night mission went out, everyone tended to hang around the radios, just in case.

Don Avila and Danny H were among the crowd waiting out the mission in the radio shack. Dennis Winn was there as well. Night missions were a unit thing.

"Do they have gunships with them?" Don asked nobody in particular.

"No," Johnnie Kane answered out of the darkness. "too hard to pinpoint the friendlies if they gotta' make a run."

"Quick in, quicker out," One of the pilots said to the night sky as they waited.

"Makin Taxi Victor, Dustoff Five Zero Two."

"Five Zero Two, Taxi Victor, Go."

"We have your lights, Ready your wounded."

We were in and out of the LZ in three minutes. Stu Willard put us down next to the wounded we got them aboard and were back at three thousand feet when Harry started checking the litter wounded, Taylor right next to him."

"Plasma Hotel, Dustoff Five Zero Two."

"Zero Two, Plasma Hotel, Go."

"I am inbound, with three US Gunshot wounds, ETA Two Zero."

"Five Zero Two, Plasma Hotel, Roger."

Plasma Hotel was the call sign for the 85th Evac Hospital, new to Phu Bai, located on the airfield.

"See that tag?" Harry asked. "That means this man has been given a shot of morphine. No more for him." Harry scuttled past the unconscious soldier and settled beside litter number two. This soldier was awake, alert, with a large bandage around his shoulder and chest. Harry gently peeled back the bandage so Taylor could see.

"Gunshot wound, through and through. Clean, not much bleeding. You're a lucky guy," Harry said to the wounded man. "You're gonna' be fine. Gold ticket home."

He let Taylor reset the dressing and moved to the ambulatory patient who was sitting peacefully in the middle of the bay, smiling. There were no bandages visible, no signs of bleeding, just a sort of happy looking guy sitting cross-legged staring out into the night.

"Where are you hit?' Harry asked. No answer. The guy never stopped staring out the door, smiling. Harry clapped his hands in the guy's face. He turned his head, his expression blank.

"Deaf as a flounder," Harry said, "Concussion probably. You see any bleeding?"

Taylor checked the soldier's ears, neck, hairline and shook his head.

"Good hit. Maybe it wears off, maybe not. Nothing we can do here." Harry patted the soldier on the shoulder, gave him a thumbs up. The soldier smiled, his mind miles away. Hopefully it would come back.

Within an hour the ship was refueled and back in the revetment. The crowd in the radio shack had dispersed, the radios were quiet and Taylor and Tighe, Don and Danny H, even Winn, weren't New Guys anymore.

Somehow, teenagers like Taylor and Tighe, like me and Harry and Woody and John and all the rest, adjusted to the madness. Bombs, explosions, bullets in the air, all induce fear and panic at a very fundamental level. Gradually, somehow, we got used to the terror. We didn't not flinch or duck or hide when the bombs went off, but we didn't go to pieces either. I managed to stay as calm and focused as the guys around me. If they hung on, I hung on. If they broke and ran, I would have been right up front with the greyhounds. So far, nobody had panicked, except on the inside. And that's how our job got done.

"Forgive yourself, you are not perfect,
Show yourself grace, you are still learning,
show yourself patience, you are on a journey."

Shannon Yvette Tanner

Chapter Twelve

Absolution

"Oh Daddy, I'm so, so sorry!" Margaret Mary wailed as she crumpled into her father's arms in the Denver hospital.

"No, no, no, my daughter, my loving beautiful child." Sean fought back tears of his own as he cradled Margaret Mary.

She had been taken from the locked dormitory of Tuloc Meadow to the hospital in Denver by ambulance two days ago. Sean had been notified the next day and flew out from Chicago on the first flight he could get. Father and daughter clung together now, each seeking absolution for sins they did not commit.

Margaret Mary continued to weep deep, heavy sobs into her father's chest. She gasped for air only to choke further on her sorrow and shame. As gently as he could Sean eased her away from him so that he could see her face, look her in the eye.

"Daughter, listen to me now, please, listen," Sean implored. With the greatest of effort Margaret Mary struggled to catch her breath and look through her tear-swollen eyes at her father.

"I love you," he began, "All of the terrible things that happened to you are over. You are safe now. I am not going to leave you. As soon as you are ready, we are going home. Everything is going to be alright. We shall get past this:

"We only begin to live when we conceive life as tragedy."

"That's Byron," Margaret Mary snuffled, looking her father in the eye.

"Yeats," Sean corrected. "And true wisdom all the same."

They hugged again, father and daughter, long and hard, as a tiny bit of the guilt, a tiny bit of the shame, and the first flickers of self-forgiveness died and grew anew within her.

"Gianoulos, pack your gear. You're going home."

Teddy peeked out from under the jeep he was working on. For the last month he had been assigned to the battalion motor pool in Danang. What he saw from under the jeep was a grinning Master Sergeant Kunstler holding a pack of papers, papers every Marine hoped he'd live long enough to see. This was Teddy's day.

Teddy popped to his feet and grinned back at Master Sergeant Kunstler.

"When, where, how and why? My count says I've got three weeks left," Teddy asked.

"You're getting a drop, because they want to pin another medal on you. You're going to Okinawa for an awards ceremony, which you deserve. Then you are going to MCRD San Diego to meet another Sergeant Gianoulos who happens to be an acquaintance of mine."

"You know my brother, Chris?" Teddy replied.

"He served under me while he was here, good man. It must run in the family." Sergeant Kunstler held out his hand. Teddy shook it.

"You said another Sergeant Gianoulos."

"You are now Sergeant Gianoulos, which you deserve as well. Congratulations, go get yourself into a proper uniform and report to the orderly room ASAP, if not sooner. With a little luck I can get you on a plane out of here today."

Teddy's head was spinning. An hour ago he was trying to break loose a lug nut under a jeep. Now he was watching a PX tailor sew sergeant's stripes onto his uniforms. He already cleared finance and personnel, got a shave, haircut and shower and said goodbye to the Marines he had met during his short stay in Danang. Part of Teddy was thrilled, part was apprehensive, the rest was dazed and confused. I'm going home! I made it! Chris was waiting to see him in San Diego, and, oh yeah, there was the whole medal thing. Too much information, too fast. Teddy couldn't decide whether to laugh, cry or throw his head back and howl. The tailor handed him back his freshly pressed, sergeant striped, ticket to home shirt and before he could put it on Teddy had to turn away and choke back a flood of tears of joy, relief and gratitude. Pulling himself together with a deep Marine Corps breath he strode into Master Sergeant Kunstler's office, with a huge smile and slightly red, swollen eyes.

"Sergeant Gianoulos, you are looking very squared away, are you ready to get out of here?" Sergeant Kunstler handed Teddy a sheaf of paperwork.

"1400 hours you are on a flight to Okinawa. Report to Flight Ops at the airbase. My driver will take you."

Teddy opened his mouth to speak, felt his throat slam shut and a softball sized lump choke him. The tears came back to his slightly reddish eyes, and he stammered, "Thank you, Sergeant."

"No, thank you, Sergeant. You have a lot to be proud of during your tour here, don't ever forget it." Sergeant Kunstler stood up from behind his desk and held out his hand once again.

"It's been an Honor and a Privilege Ted. Save journey to you, Semper Fi."

Teddy shook his hand, turned and left the office, his eyes much more than slightly red. Three days later Teddy stood at attention on the Parade Deck in Okinawa. Teddy was to receive the Bronze Star for Valor, his second award of the medal, along with a Purple Heart for the shrapnel he had in his leg.

Two hundred Marines were in formation behind him, also at attention, as a full bird colonel read,

"Attention to Orders:

Sergeant Theodorus Gianoulos is awarded the Bronze Star for valor. His citation reads, "For Uncommon acts of valor while under direct enemy fire during ground operations against a superior hostile force in the

Republic of Vietnam, Corporal Gianoulos distinguished himself by his heroic actions while serving as a rifleman for Alpha Company on 12 February 1968.

That day Alpha Company came into contact with a superior enemy force in the hills surrounding Con Thien Fire Base. The company established a defensive position on the hillside and began to receive heavy mortar fire. Corporal Gianoulos determined the location of the enemy mortar tube that was viciously pounding the platoon positions of Alpha Company. With total disregard for his own safety and braving intense enemy gunfire he maneuvered into a position to use his weapon to fire down upon the enemy mortar site. The enemy gunners were both shot, which immediately knocked the enemy mortar out of service.

Enemy soldiers then attempted to overrun Sergeant Gianoulos' position and, though wounded himself, he held this position killing several attacking soldiers. Sergeant Gianoulos then refused evacuation until enemy contact was broken and his own squad was secure. His actions stopped further enemy activity, reducing casualties and saving lives. His bravery and professionalism is in keeping with the highest traditions of military service and reflects great credit upon himself, his unit and the United States Marine Corps.

Teddy sharply returned the salute of the Colonel and watched as the formation of Marines marched past him in review. That old softball lump had returned but three hours from now, he was boarding a plane for San Diego. Teddy's war was over, at least for now.

The basilica at the Catholic University of America in Washington, D.C. rose majestically around and above seminarian Peter Rayburn. Peter was finishing his second year of preparation for the priesthood. He found his studies daily brought him closer to his vocation, and more self-assured in his choice to become a Catholic priest. As he knelt, enjoying his few solitary moments of meditation and prayer, a student disciple approached him.

"Peter," he whispered, "you have a telephone call in the deacon's office. You should come right away."

When the deacon handed Peter the telephone his look of concern and sympathy transformed to one of fear and anxiety in Peter. Phone calls to seminarians were never casual, most often heralding bad news. When Peter said hello, the voice he heard was not one he had expected,

"Peter, this is Sean Sullivan, Margaret Mary's father."

Peter felt an initial wave of relief as the score of family tragedies he had conceived on the way to the office faded. But that anxiety was replaced with concern as Sean continued, "My beautiful daughter is in pain, a terrible thing has happened to her and it's my prayer that you can help."

Peter listened to the story, fierce anger rose within him, replaced by deep sympathy, perhaps an offshoot of his seminary training.

"How can I help?" He asked.

"Could you find it within yourself to come here? We are at my sister's home in Hammond, Indiana. I will make any travel arrangements you require."

Peter looked up to see the deacon nodding his head in approval. Sean had already explained the reason for his call to him.

"I'll be there as soon as I can. Peter replied.

Sean crossed the busy terminal at O'Hare airport in Chicago. Peter had deplaned and was waiting by the luggage carousel. They spotted each other and after a brief wait for luggage were on their way to Rose Sullivan's home.

"So how is she? Has she gotten any better?" Peter asked.

"She barely eats, she sits in her room, we can hear her crying," Rose replied as she weaved through the airport traffic.

"The only thing she has asked for was you or David or Teddy to come to her. David and Teddy are still in Vietnam. She did not know that, and the news made her feel worse. That is why I called you and why it is so important that you are here. She needs to remember who she was before all this happened to her."

"I'll do whatever I can," Peter replied, praying that he would know what the right thing was.

"And after that I felt like whatever he did to me, I deserved." Margaret Mary spoke in a hush, staring only at her feet as she and Peter walked around the block from Aunt Rose's home.

Peter stopped, tugging Margaret Mary to a halt with him. As gently as he could he put his arms around her. Margaret Mary cringed, trying to make herself smaller, trying to squirm away. Peter let go, stepped back, holding tight to Margaret Mary's hands.

"Do you remember what an awful pain in the ass I was in school, always waving my hand around and asking, "Why?"

Margaret Mary couldn't help the wisp of a smile as she remembered Peter's innumerable trips to the cloakroom purgatory.

"All those times I was always sure I was asking the right question. I felt I was right. One of the hardest things I've had to learn at the seminary so far is that feelings are not facts. Just because I believed I was right didn't mean it was true."

Margaret Mary looked up from the ground. She tried to pull her hands loose from Peter, then stopped. "Peter, I am very happy that you have found happiness in the church. I have lost mine. Lost it and I don't think I will ever get it back."

"And for that you feel very guilty, right?" Peter asked as kindly as he could.

Margaret Mary could only shake her head Yes and begin, once again, to cry.

"Feelings," Peter whispered, putting his arms around Margaret Mary, "are not facts."

Which led to a renewed agony of sobs from Margaret Mary. This time, however, she did not try to pull away from Peter, but leaned into him shaking from head to foot. Peter stood very still.

Finally, the sobbing stopped and the shaking shook less. Margaret Mary raised her head from Peter's chest and, in a very low voice whispered, "Thank You."

Before Peter answered he motioned to a park bench. They sat close to one another, as friends will. Then Peter said, "Thank me by answering this question, as honestly as you can. I promise I will never repeat your answer to anyone."

Margaret Mary nodded, and Peter continued, "What, exactly, is it you feel so guilty about?"

Margaret Mary started to reply, stopped as a wave of deep shame washed over her. Unable to look Peter in the eye she said, "I slept with Kendall, we had sex many times. He did not force me, I wanted to."

"And why did you sleep with and have sex with Kendall?" Peter asked gently.

"I loved him, or at least I thought I did. He told me having sex would bring us closer. I wanted to be closer."

"And were you?

"Yes, at first. He was so gentle and kind. I really thought I loved him, I did. When we got to Colorado he changed. He wanted to sleep with the other women there. He wanted me to sleep with the other men. I refused and…"

First there was a deep hiccup, then the tears started again. Margaret Mary hid her face in her hands. Peter sat quietly, waiting. Eventually she looked up, eyes red and swollen.
"He hit me, hard. In front of the others, told me I was worthless and ugly. He told the others to shun me as well, and they did."

Anger was replacing her tears as she recalled,

"I tried to run away, twice. I got lost in the woods and Kendall came after me. He dragged me back and said if I ever did that again he'd leave me out there to die. I tried again, the very next day and this time when he caught me, he beat me. I was forbidden to leave the dormitory; they even tied my ankle to a bed. When the snows came and winter set in, I knew I could not get away. I was barely alive when the police came. Sometimes I wish they hadn't come at all, and I just died there."

Margaret Mary stopped talking, awash in humiliation and hopelessness. Peter seethed, himself overwhelmed with very un-priestly feelings toward Kendall and the rest of the Tuloc Meadow tribe. He struggled with the anger, let it pass and finally said, "My heart is truly, deeply broken that this has happened to you. I can only remind you that you did nothing to deserve this treatment. What you did, sleeping with Kendall, having sexual relations with Kendall came from a feeling of love within you and you acted upon it. Is that true?"

Margaret Mary nodded.

"And Kendall treated you well and with love as you fell in love with him?"

Another affirmative from Margaret Mary.

"The evil that came afterward, that came from Kendall who you felt you were in love with?"

"I was so stupid!" Margaret Mary pounded her hand on the bench.

"Were you stupid, or were you trusting? And did you have any reason for not being trusting before this terrible evil befell you?"

"I thought I loved him. I thought he was a good person. I thought he loved me."

"And are all of your thoughts perfect Margaret Mary Sullivan?"

Margaret Mary's celebrated look of practiced patience and forbearance changed to a flash of anger and then exasperation.

"I do not think I am perfect!", she snapped.

"Neither do I", Peter chuckled. "You made a mistake Mags, you trusted, and you loved someone who you thought loved you. It happens, we are not perfect, not even close, and you, of all people are entitled to a misjudgment based on love as much as anybody."

"I've never done anything that stupid before," She replied with an irritated sniffle.

"Welcome to the human race," Peter laughed. "I just joined myself, so welcome aboard."

Which got the first smile Margaret Mary had allowed herself in a long, long time.

"Would you do one more thing with me? It's an exercise we all had to do in the seminary. It really helped me."

Margaret Mary nodded, and they faced each other on the bench. Peter once again took Margaret Mary's hands.

"Am I right in remembering that Linda Lewis was your closest girlfriend in high school?"

"Until she moved away," Margaret Mary answered.

"Yes, moved away, we'll get to that. But she was your best girlfriend for a time, wasn't she?"

When Margaret Mary agreed, Peter continued, "Imagine, if you will, that everything that happened to you, everything, happened to Linda and that she was feeling as terribly guilty and lost as you are. What would you tell her?"

Margaret Mary gave this question the deep thought it deserved and chose her answer carefully, "I would tell her it wasn't all her fault and that she should forgive herself."

"As in, "To err is human, to forgive divine."

"That's Shakespeare, right?" Margaret Mary hearkened back, as the Irish often will, to the words of The Bard.

"No," Peter laughed, "That's Alexander Pope."

"I should have known that." Margaret Mary chided herself, "I'm out of practice, with my reading."

"Who is not important. What it means is very important. From it I learned that forgiveness is more essential than guilt." Peter lapsed into full, almost a priest mode.

"I can't forgive him, Peter, I can't. I hate him."

"Not him, Mags, yourself. You need to forgive yourself. You deserve it as much as Linda Lewis would have."

And within Margaret Mary's heart a light flickered back on. This light would have to be nurtured, reinforced, encouraged and brightened but now, at last, it was shining in the darkness. "I'll try," she said, "But I don't know if I can."

"Nobody does, until they try. You have always been the brightest, smartest person I know, Margaret Mary. I feel very strongly that you have it within you to do this bright and smart forgiving thing."

Margaret Mary scooted across the bench to hug Peter. He hugged her back and then straightened, "There is one more thing," Peter added, "I saved it for last because it is for me as much as it is for you." Peter removed an envelope from his pocket. He carefully removed the letter it contained and held it out before him. "Can I read you this? It's from Dave, in Vietnam."

Margaret Mary felt her breath catch and her eyes went wide with surprise. She couldn't answer, only shake her head, "Yes" and stare, transfixed at the pages. Peter began reading,

Hello Peter,

Thank you for writing. It means a lot to hear from you as our old gang has sort of scattered to the winds. And the baseball cards! I almost forgot there was any such thing as baseball cards! I can't believe I missed the whole World Series, and our Red Sox! That must have been so much fun. I showed them to all the guys, they were like reminders of another world. I am in a place called Phu Bai which is just

south of the Demilitarized Zone between North and South Vietnam. I don't know why they call it a demilitarized zone; an awful lot of people are shooting at each other up here. My unit, I am very proud to say, is a helicopter ambulance unit. We pick up the wounded from the battlefield and get them to field hospitals. It is noble work Peter, and I am proud to be part of it.

Has anyone heard from Margaret Mary? I miss her a lot and hope she is doing well. Please let me know if you hear any news. I know Teddy is here, in Nam, but I don't know where. The Marines move around a lot. I've written to him but no answer. My Mom and Dad tell me that Connie lets them know that he is alright whenever they get mail. I sure hope so. When you left to go be a priest and Margaret Mary left for school and Teddy joined the Marines, I felt like I was all alone on the moon. You guys had all chosen to go do something, I was lost, kind of. I don't feel that way now. I am with the best people I have ever known, Peter. Best besides you guys of course. The guys I am with risk their lives every day to save people, and several of my buddies have been seriously wounded doing so. They are the greatest. I am proud to be one of them This place has changed me. Before I got here I used to think about all the good things that were happening, or the good things that were ahead for all of us. Here, I have to be aware at all times of the bad things that are going on around me. Once I stopped being terrified, I just became kind of numb. I can't let myself get too happy or too sad. I just have to stay alert, ready, for whatever happens. Lowell, Margaret Mary, Teddy and you all feel like a thousand years ago to me. I don't know if I can ever get back to being who I was before, or if I want to. I miss all of you, but not too much, I can't afford to. Please let me know if you get any news about Margaret Mary. I know it's no secret how much I care for her. I hope she is okay and happy, I wish I could say that I was. Anyway, thank you again for your letter and the baseball cards. PLEASE write to me whenever you can, I need to know there is somewhere else beside here and that out there you and Margaret Mary and Teddy and I can maybe get together again.

Your Friend, DF

When Peter finished reading, he handed the letter to Margaret Mary. "His address is on there if you want to write to him," He said, hopefully.

Margaret Mary stared long and hard at the letter. She let her emotions wash over her, guilt and sadness, love and affection, happiness and misery, anger and compassion, each in a wave, cleaning and cleansing. Then she smiled. "Of course, I will write. Today. Thank you for sharing this with me, Peter."

And they walked home together, holding hands.

Chapter Thirteen

Attention To Orders

"DEPARTMENT OF THE ARMY UNITED STATES ARMY OF THE REPUBLIC OF VIETNAM SPECIAL ORDER 262 DATED 11 NOVEMBER 1968 HEREBY AUTHORIZES MAJOR DAVID DUARANT TO ASSUME COMMAND OF THE 571ST MEDICAL DETACHMENT EFFECTIVE THIS DATE."

We were standing at attention in unit formation as the change of command ceremony took place. This made it official, the first-year tour of the 571st was coming to an end and those who had arrived twelve months ago would be going home, sooner, much sooner, than later. The ceremony marked the transfer of leadership from Major Vince Cronin to our new commander Major David Durant. Two finer men I would never know.

"Are you sure about this young man?" Major Cronin asked as I stood in front of his desk with Woody, Harry and John Kane the day before the change of command ceremony.

"I am Sir, I've given it a lot of thought and we have talked about this among ourselves."

Woody, Harry and John had asked me to be the spokesman for our group. We decided we were going to extend our tours another six months.

"Can you tell me why you decided to do this?"

"I've got less than two years left on my enlistment sir, so do the rest of the guys. That means we'll spend the rest of our time stateside being a clerk or an instructor a repairman, but no matter where we get assigned, it could never be as important as what we do here, sir. I want what I do to matter, this is where I feel like it does."

"And is that how the rest of you men feel?" Major Cronin asked.

Woody, Harry and John all answered, "Yes Sir".

"I don't have to tell you about how dangerous it is here," Major Cronin continued, "Extending your tour extends your exposure to that danger. Have you thought carefully about that?"

"We have sir." The others nodded as I spoke.

"Then I will pass my recommendation on to Major Durant. It will be his decision whether or not you will stay in this unit."

"Sir," Woody piped in, "I don't want to serve in no other unit, I want to stay in this one."

"Then I suggest each of you get into proper uniform before seeing Major Durant and take that up with him."

"Sir," John Kane spoke up, "this is the cleanest uniform I have; the rest are in the laundry."

We all wore standard jungle fatigues, faded a stylish light green and adorned with our rank and unit patches. These were our "show off" uniforms, not the ones we worked in every day.

"Well then I suggest you read these," Major Cronin passed us each a mimeographed set of orders, "and dress accordingly."

We read the papers. They were promotion orders. We were, each and every one of us, promoted to E-5, the highest rank we could hope to achieve as first-time enlistees. It was an honor and we knew it. Major Cronin smiled.

"You men have done a good job during our year here. Specialist Giles, you have made some very unwise and immature decisions prior to this, but you have been an excellent medic. You are thought highly of by your crewmates and the doctors up at the surgical hospital as well as by myself. Congratulations. Woody, you and John have done your jobs as well as I could have expected them to be done and more. Thank you for that. Young man, I have not forgotten that when I called for volunteers to be patient protectors on our ships yours was the first hand to go up, along with Woody and John. You all deserve this promotion, and I am proud to have served with you."

Which was about as emotional as four E-5's and one Major can get without getting arrested or medicated. We all shook hands with the man who made us men.

The next morning the whole unit assembled on the helipad. A substantial crowd of doctors, nurses and hospital personnel were also there. Two of our ships were cranked up, loaded down with the eight guys who were rotating home today. They would be flown to Da Nang, out processed and forwarded to Cam Rahn Bay for the jet ride home.

Major Cronin was one of them, Dennis Hughes and Dwight Anderson were two more. Our practically useless but thankfully invisible first sergeant Willie Johnson another.

The landing skids of both ships were wreathed with colorful smoke canisters which would be popped during their fly-by over the pad.

Fate had been kind to us, very kind during our year in Vietnam. We had several guys wounded, lots of ships shot out from under us, countless mortar and rocket attacks, one deadly snake bite, two vehicle mishaps and several broken hearts. But nobody had been killed.

In the twelve months following our arrival over 6,700 U.S. soldiers were Killed In Action. Over 30,000 were wounded. South Vietnamese military and civilian casualties were over 100,000 killed and wounded. Enemy casualties in North and South Vietnam also ran into the six figures. The carnage had, for the most part, passed us by.

The 571st Medical Detachment flew an average of four rescue missions a day, every day. In excess of 1200 missions carrying three wounded soldiers per, resulting in a total of approximately 4,000 wounded ferried safely to hospitals.

The bulk of the men who carried out our mission were going home. Woody, Harry, John and I knew we did the right thing by extending.

Major David Durant snapped a hand salute as the ships lifted off, circled the pad and then flew low and fast over the crowd, smoke streaming from the skids. Joyous, happy faces smiled down at us as we said goodbye to our comrades.

"Each of you will receive a thirty-day leave. I will expect you back here at the end of your leave prepared to do the same fine job you have done up to this point. Any questions?"

Major Durant, now in command of the 571st was briefing the four of us on our extension protocols. We would be going back to the States. We would be getting a lot of inoculations. We would have one week travel time in addition to our thirty days leave. We should conduct ourselves as a credit to our unit. Specialist Giles must not marry, re-marry or get engaged. We should pack and get going.

"There is one more thing," Major Durant said. "When you return, I will expect you to act as role models for the new men. They will look to you as experienced, be sure to demonstrate the behaviors they will need to survive this tour."

All four of us answered in the affirmative, saluted and left Major Durant's office. Role models. More new ground.

PAN AMERICAN

Chapter Fourteen

Homeward Found

The three days travel from Phu Bai to Fort Lewis, Washington was a blur of excitement, anticipation, duplicated out and in processing, no sleep, lousy food and long, long plane rides.

Finally stateside and grinning, Harry caught a flight to Philadelphia and to God knows what domestic wrath awaited him there, Woody to Pittsburgh, the closest airplanes actually got to Parsons, West Virginia, John Kane to Los Angeles and me to Boston. We promised to meet back at Fort Lewis in thirty days.

My flight left last. Sitting in Seattle airport dressed in a short sleeved, tropical khaki uniform which looked entirely out of place in wintry Washington state, I didn't notice the stares, at first. Then it was all I could notice. It was as if there was a bubble around me. Old folks peeked and looked away, younger folks glared or gawked. I noticed the other service men and women scattered around the airport were receiving the same treatment. Nobody was rude, nobody was friendly. I felt invisible.

I sat watching the crowd flow by. Younger guys were wearing their hair longer. The girls' dresses were shorter.

Hurray. Everybody seemed to be in a hurry. Nobody was smiling. Welcome to America in late 1968.

My flight boarded on time, and I had a three hour layover in Chicago. O'Hare airport was about the same as Seattle, bigger, busier and much colder. My uniform was becoming less stylish and practical the further east I travelled.

"You look lost." A very pretty, pink and green uniformed stewardess said as she sat next to me.

"Dazzled is more like it," I answered. "I've been away for a while, everything looks different."

"Vietnam," She stated. Stewardesses were very good at reading service uniforms. I nodded, she continued, "Was it terrible over there?"

"Not all the time," I said, wishing the subject would change.

"I'm Kathy," she said holding out her hand.

"Dave, nice to meet you." Nice hand, soft.

"Welcome home. Where are you headed?"

"Boston, first, then home to Lowell."

"Aren't you cold in that uniform?"

"I wasn't in Saigon, am now though."

"Come with me," she said, holding out her hand and standing up. I had two more hours till my flight. I went with her.

We passed through several secret corridors, past barriers that said "Authorized Personnel Only" and into a room filled with lost luggage, hats, coats, sweaters and paraphernalia I could only guess at. A man sitting in a wire caged area looked up from his newspaper.

"Hi Kathy, find another one?"

Kathy smiled, waved and guided me to a rack of coats, overcoats, sweaters, and jackets. She considered me for a moment, sliding hangers until she found what she was looking for.

"It's Air Force, but no one is going to know the difference. Try it on." She held a dress blue Air Force issue overcoat out to me. Fit like a glove.

"I can have this?

"Only if you buy me a cup of tea. What do you think, Sam?" Kathy asked the man in the cage.

"Been here about a month, nobody asking for it." Sam went back to his newspaper.

"Maybe we should remove these," Kathy said, unclipping a pair of full-bird, colonel wings from each topcoat shoulder. She took a step back and regarded me with an appraising eye, "Handsome, you are going to slay the ladies."

Slaying wasn't what I had in mind. "Thank you," was all I could mutter as I discovered I could still blush.

Kathy was from Dayton, Ohio. She had worked for the airlines for three years. She was especially fond of servicemen because her brother was in the Navy. She drank a brand of tea I had never heard of, chamomile? I drank Tetley. Kathy brought me up to date on many things including the odd stares I was getting in the airports. The country was divided, she explained, pro-war and anti-war. The division was generational, economic, racial, political and just about everything else. Uniforms made most people uncomfortable. Rock & Roll was mostly peace, love and dope. Elvis was a mannequin. Nixon was going to be President. It was shaping up to be an interesting thirty day leave.

"So, do you have a girl back home?" Kathy asked as we finished our tea.

"Not anymore, used to." I was kind of wondering about the "girl back home" thing. I hadn't heard a word from Margaret Mary in over a year, there was a long-ago Becca, and a Susan, who had actually written to me twice while I was in-country. I hadn't written back.

"Well, you will," Kathy laughed, "Some lucky girl is going to scoop you up the minute they see you in that snappy blue raincoat."

I could only hope so. Kathy and I parted ways an hour later. She gave me a hug and a ray of hope and kindness. I hoped the rest of the guys were being as lucky as I was.

Four hours later I was looking out the window as we descended into Logan airport. There was a tightness in my chest, and I couldn't seem to catch my breath. I didn't know what to say, I didn't know how to act. I didn't know how to be. When the plane landed, I stood up, took a deep breath and walked back into the world.

I could see my Mom and Dad through the glass wall of the terminal. They looked older. My brothers were there, they looked taller. Behind them, in full United States Marine Corps dress green uniform was Teddy! Everybody was smiling. I was fighting back tears, good ones.

My Mother hugged me so hard I thought she would never let go. My Dad beamed and patted my shoulder, his eyes looking awfully wet. My brothers, John and Bob, looked happy as well, ogling my uniform and my snappy new raincoat.

Teddy stood back at first while we got the family thing out of the way. Then he stepped forward with a huge grin and held his hand out for me to shake.

"Welcome home, Dave, glad you made it." I stepped around his hand and gave him a military bear hug.

There was a luggage carousel, a wintry parking lot and shortly thereafter I was in the front seat of the family car humming along Route 93 to Lowell. My Dad seemed to be driving awfully fast, way above jeep speed. It was quiet inside. I was going home, I hoped it still felt like home.

"How could you do this to me! How could you do this to our baby! You are such an asshole!"

Harry was cringing in an overstuffed chair in his parents' living room as Linda screamed at him. His mother stood nearby, arms crossed and scowling while his father peeked in from the kitchen every now and then, a secret smile on his face, out of the bullseye. Harry's sister, Karen, stood nearby ready to hand Linda any convenient throwing objects. Harry could have had a more peaceful leave if he had gone to North Vietnam.

Woody woke up on his first morning home and found he didn't want to go hunting, something he had enjoyed all his life. The trip had been planned by his Dad, his uncle and best friend. Woody hesitated in his bed, wondering how to tell them he had enough of guns for a while.

Johnnie Kane raced his beloved Harley Davidson Knucklehead down the main street of his hometown in Santa Ana, California. He was surprised at how small everything looked, how differently he saw things that were exactly the same as when he had left. Only smaller.

"They're paying me to punch lieutenants!" Teddy laughed as we sat in Lefty's sharing coffee, English Muffins and war stories.

Teddy had been back in the World for a little over five months now. He was stationed in Quantico, Virginia as a boxing instructor to the Officer's Candidate School classes. He took leave when told I was coming home.

"Perfect!" I answered, "How do I get into that line of work?"

"Wrong branch of the service man, Birdie on the Ball, the only way to go." Teddy was thriving in the Corps. He was a decorated combat veteran, with a deluxe assignment, who spent his days punching officers.

We were enjoying ourselves, catching up, but there were some serious issues to discuss.

"You got wounded. How bad is it?" I knew about wounded, there were no good wounds.

"Shrapnel, hurt like a bastard at first, mostly in my legs. I healed up really well, Golden Ticket and all that." Teddy poked his coffee cup. This was not an easy subject for either one of us.

"You? You make it alright?" He asked.

"Not a scratch. You guys were doing all the heavy lifting." I was remembering the gruesome days of Tet, the battle for Hue Citadel, Khe Sahn and Con Thien, Marine Corps battles, every one a victory.

"Still man, Dustoff, you guys were the best." Teddy answered.

"Still are, Ted. I'm going back."

"You're what?" Teddy wasn't only surprised he was shocked.

"I extended, volunteered for another tour. I've got thirty days leave then back to Phu Bai."

"Do your parents know yet?" He asked.

"Haven't told 'em. I'm not looking forward to that either."
"Your mother is going to freak out."

"Yeah, I think we can count on that."

"Hey, Peter's coming home this Friday. He really wants to see you," Teddy mentioned, artfully changing the subject.

"He wrote to me a couple times in Nam," I answered, "It was good to hear from him."

"He wrote to me too," Teddy said, "Sent me a prayer book."

"What about Margaret Mary?" I finally asked, she had been the great silence, the elephant in the room since I had been back.

Teddy grew solemn, "There's something you gotta' know. Don't get mad alright?"

"Okay," I answered while I was getting mad.

"Some guy beat her up. Bad. She was in the hospital. She's out now and staying with Sean and her Aunt Rose in Chicago."

Teddy's words hit me like stones, big stones.

"How? Why? When?" Was all I could stammer as my anger transformed into distress.

"Some guy named Kendall beat her up. He's in jail now. Peter told me all about it."

Kendall. My anger made a raging comeback, but I struggled to concentrate on my concern.

"Peter told you about it?"

"Yeah, he went out to see her in Chicago. He wants to talk to you before you contact her, asked me to ask you."

"He'll be home Friday?"

"Day after tomorrow. I think you should wait."

"She's okay now?"

"Better, Peter knows a lot more than I do."

Everybody seemed to know a lot more than I did. It was going to be a long wait till Friday.

"Can we all sit down for a minute? I need to tell you guys something." This was a talk I wasn't looking forward to having. We just finished dinner, John, Bob, Mom, Dad and I. We were about to scatter for the evening, but I had to get this off my chest. Everybody looked at me, hints of worry and curiosity as they settled back down at the dinner table.

"I'm going back to Vietnam after my leave is over. I volunteered for another tour."

The kitchen got quiet. Eyes got wider, questions started, then stopped. My brothers fidgeted, my Mom exploded, "Why! What did you do that for! Do you have any idea how much we worried about you for the past year! What you put us through! And now you're going to do it again!"

She wasn't finished, just out of breath when my father asked, more gently, "Can you tell us why you decided this, son?" I leaned forward on the table and tried really hard to put my reasons into words.

"My unit, the guys I'm with over there, they are the best people I have ever known. Our job is to get wounded people to hospitals. Sometimes, most times, it is kind of dangerous, but they don't care, they go get them anyway. We save lives, every day. I'm very proud to be one of those guys. I feel like I'm doing the right thing and I want to keep doing that."

"You did your share! You were there a whole year! Let somebody else get killed doing the right thing!" That was it for Mom, she stormed out of the room leaving a black cloud of anger behind her. Nobody followed.

"Dad, when you were with your bomber crew during World War II you were friends with those guys, right?" He said he was.

"Were you proud to be with them?" He nodded.

"That's how I feel Dad. I'm proud to be with them and I don't want to leave them behind."

"We fought Hitler, son, and the Japanese after that. I'm not sure who we are fighting or why over there. Are we doing any good?"

"I don't know about "we' Dad, but my unit is. We save lives, every day."

"Then I guess you should keep doing that. I'll talk to your mother."

Good luck with that.

"That's the way she wants it, Dave. That's what she asked me to tell you." Peter was half explaining, half pleading for me to understand.

Margaret Mary was in recovery, seeing a therapist twice a week. She had started back to school in September, taking two courses and trying to get back on her feet again. Peter explained that the Denver police wanted her to return to Colorado to testify against Kendall, but she was reluctant to go. Peter had seen her only that one time and she asked for some time to mend before seeing anyone. I was one of those "anyone's". I could call, she wanted to hear my voice, but she was not ready to see me, or anyone else from before, before meaning before Kendall. I was having my own very dark thoughts about Kendall.

Meanwhile, as Christmas drew nearer, my Mother's anger at me was such that she wouldn't speak to me and would hardly stay in the same room with me. The whole house was infected with her fury which seemed to increase with every passing day.

Teddy returned to Quantico and Lowell seemed empty and sad. No Ho-Ho-Ho this year.

Peter arranged for me to call Margaret Mary on Christmas Eve. When the time came I was so nervous I was relieved we weren't having this meeting in person. I could only imagine how hard this must be for her.

"Oh David, I'm so ashamed," Her voice wailed as she broke my heart. I felt I had let her down or that this was somehow my fault, something I could have prevented and must avenge. I couldn't find words to respond. I was literally choking on my own emotions.

"I was such a fool! Can you ever forgive me?"

It never occurred to me that there was anything Margaret Mary needed to be forgiven for in this whole horrible situation. I felt I needed forgiveness for not being there to protect her.

"But I am so glad that you and Teddy are home safe and don't have to go back to that terrible place."

Just when you think a conversation can't get any worse.

The conversation with Margaret Mary didn't end well. I asked, no begged her to let me come see her. She said she needed more time. Also, she was shocked, even angry that I was going back to Vietnam. It seemed everyone had an opinion about this and none of them were good. Volunteering for a second tour didn't make a lot of sense to many people, to any people actually, but it made more than sense to me. It was what I decided was the right thing to do. It was my decision, alone, without permission or assistance from anyone. Returning felt like the right thing to do and I was going to stick by that. Everyone else was going to have to adjust.

While I was very disappointed that I wasn't going to Chicago to see Margaret Mary, I was, however, going to Philadelphia.

Harry called me the day after a very bleak and bleary Christmas. The gloom and anger in my home on Glenmere Street had not subsided. My Mother was still on full warpath, my Father and brothers cowed by her wrath. I was the prodigal son unforgiven, at least by Mom. No one, in fact, even Teddy, thought my returning to Vietnam was a good idea. No one except me.

To make matters even more complicated I had slept with none other than Donna Delancey twice over the holidays. No "pretend sex", high school hi-jinks this time. Full on, drunken, baby making boom-boom. I had yet another item to add to my "What was I thinking list?" And then Harry called.

"What do you mean you're not going back?" I stammered into the phone after Harry announced he was going to Alaska with Linda.

"Canada doesn't make you go back if the cops come to get you," Harry announced, "Lots of guys go there to get out of the Army."

I barely knew where to begin, or even if I should. "Harry, you do know that Alaska is not part of Canada, don't you?"

"It isn't?"

Since my thirty day leave in Lowell had run its course I decided to fly back to Fort Lewis by way of Philadelphia to see if I could talk some sense into the best medic in our unit.

My farewells in Lowell were brief and somewhat painful. My Mother was still not talking to me. My Dad understood, sort of, but I could appreciate his concern. Margaret Mary was more sad than mad, but the combination was troubling. Teddy and Peter wished me luck. Donna Delancey wanted us to be pen pals.

Going to Philadelphia felt like a rescue mission and an escape.

"Yo, Dave, I looked it up, Alaska's right near Canada." Harry seemed reassured by this news. His plan had not changed.

He and I were sitting in his neighborhood bar named the Harrowgate Tavern. Harry was drunk. The bartender was drunk. Everybody in the place was drunk. I was getting drunk.

"If you go to Alaska, you will get caught. Then you will go to prison. Linda and your daughter will be thousands of miles from home with no money. It's a bad idea."

Harry was processing this information between shots of ginger brandy and glasses of draft beer. I was trying to dispense wisdom by doing the same thing. The outlook was bleak.

I was in Philadelphia for three days. I stayed in the basement of Harry's parents' house on a very lumpy sofa. Linda, with child, shared Harry's bedroom upstairs. We were scheduled to meet Woody at Fort Lewis in two days. I fought off the effects of three days worth of beer and ginger brandy and made one last attempt at talking some sense into Harry.

"I checked Harry, it's five years in prison for desertion. Military prison, Fort Leavenworth, Kansas. You will do every single day. And you will get caught."

Harry's family was sitting around the dining room table as I spoke. There was a lot of solemn head nodding and concern. Linda was crying, so was the baby.

"You've got, what, seven months left in the Army when we go back. You don't have to fly. Major Durant will get you transferred to the hospital, and you can work on the wards until your discharge." I turned my attention to Linda, "Phu Bai is pretty safe now. It's a big base. If Harry stays out of the helicopters, he'll be fine. I'll keep an eye on him, I promise."

Sanity was making a temporary comeback to the room. I had Harry three quarters talked into returning. His family, except for Linda, was coming around. I played my last card.

"Look, there's something you guys should know. I served with Harry for a whole year. In my opinion he's one of the best medics in our unit. That's why I'm here. I'm his friend and I don't want to see him go to prison. Harry saves lives. He needs to keep doing that."

Finally, there was consensus in the room. Harry would leave with me tomorrow. No Alaska. No Fort Leavenworth.

It was settled but there was one more anguished cry from Linda however, "What if he gets married again?"

Tomorrow couldn't come soon enough.

Chapter Fifteen

Back In The Saddle Again

Technicolor, Cinemascope & the NBC Peacock.

I know in my heart this is an unhealthy way to describe my feelings as I, along with Harry and Woody, travelled back to Vietnam. Yet there was a sense to the returning, an intensified sensation heightened by anticipation of rejoining and undertaking much larger and more vital than I ever experienced before. I was returning to a reality that was more alive, more significant and much more vital than the one I was leaving. The closer I got to Phu Bai the more intense the colors seemed, the deeper the sounds reverberated, the faster my pulse, pulsed. The nearer the danger the more I felt alive. No wonder shrinks love talking to me.

What I recall most is that at this point, The World was a black and white movie while Vietnam was in full, raging, insane Cinerama. In Vietnam I felt essential, in Lowell I existed.

"Yo Dave, isn't Fort Lewis right next to Alaska?" Harry still had the notion in his head.

"Harry, Fort Lewis is in Washington. Washington is next to Canada. Canada is next to Alaska," I answered patiently.

"That's what I said," Harry answered.

Harry and I met up with Woody at Fort Lewis and started our journey back to Phu Bai. John Kane had already returned. He stayed only two weeks of his thirty-day leave. In a brief phone call before he left, he said, "There's nuthin' here except cheeseburgers and TV, I'm goin' back."

I knew exactly how he felt. Lowell had become a drab little town full of dilapidated station wagons and pothole studded streets. My Mom and Dad labored from payday to payday, my brothers from fads to fantasies. Teddy was engrossed in his Marine Corps career in far away Virginia. Margaret Mary was in Illinois, distraught, distanced in her recovery, damaged but healing. What I saw behind me was my childhood, ahead lay my future.

"So, Woody, did you see that girl, Carol, while you were home?" I asked during the long, long plane ride across the Pacific.

"I did," Woody responded, "but you know Dave she took me to meet some of her friends and I felt like some kind of a pet she brought along."

"Were you wearing your uniform?" I asked.

"I did, I thought she would like that, but she seemed kind of embarrassed that I wore it."

"Yeah, I didn't wear mine at all after the first day, I got a lot of funny looks."

"Weren't nuthin' funny about the looks I got," Woody replied, saddened.

"Me neither, Woody. Harry still asleep?"

Woody craned his neck to spot Harry across the aisle and one row back, out like a light.

"Yeah, was he really thinkin' on not going back?"

"I don't know if thinking is the right word to describe how Harry functions."

Woody chuckled, I sighed, and the flight flew on.

Cam Rahn Bay was our entry point in Vietnam. An enormous compound of sand, barbed wire, gun towers, sandbags and gray, nondescript hooches and barracks. There were lots of GI's in jungle issue green, blast walls, bunkers and bomb craters. The atmosphere was one of great undertakings. Bulldozers bulldozed, giant cranes lifted, gunships hovered, and artillery thundered. Home sweet home. We processed in and processed out as quickly as possible and set off to hitch a ride way up north. At the airfield we bumped a trio of hapless Privates from a C-130 flight to Danang. It felt good, in an evil sort of way.

"Don't be in such a hurry guys," I advised them, "It's a lot safer here than where you're going."

A couple of bumpy air hours later we were in Da Nang.

"Get in here Ferrara-ay, and bring your two buddies with you."

I couldn't believe my ears. Command Sergeant Major Hoadwonick, formerly of Fort Jackson, South Carolina was now head of personnel at 44th Medical Brigade in Da Nang. He was also a Command Sergeant Major, the highest rank an enlisted man could achieve in the Army. A Command Sergeant Major is more powerful than a General, more powerful than the Pope, more powerful than the President of the United States, especially if you were standing in front of his desk in Da Nang. He remembered me. Not yippee.

"Are you still trying to rat fuck my Army Fernari-ah?"

"No Sergeant Major, I would never do that," I replied, standing at strict attention as Woody and Harry were wisely doing.

"What about you Wood- Man?"

"N-No Sergeant," Woody stammered.

"What is that in your mouth, Wood-Man?"

"Skoal, sir, chewing tobacco, I…"

"I know what chewing tobacco is Wood-Man. You will now exit my office, take that disgusting wad of garbage out of your mouth, dispose of it properly and hygienically and get back in here. You have thirty seconds."

Woody spun on his heel and scampered out of the office. Command Sergeant Major Hoadwonick turned his attention to Harry. There was no way this was going to go well.

"Giles, until I read your 201 file I thought Fernarier was the biggest fuck up I ever met in my Army. As it turns out, he is not. Congratulations, you are now in first place."

Harry looked confused but at least he knew enough not to say thank you. Silence is golden.

Woody thundered back into the office and snapped back to attention. Command Sergeant Major Hoadwonick eyeballed us. "Do any of you have the slightest idea how I ended up behind this desk?"

Silence was still golden.

"I came to Vietnam as a forward artillery observer. I volunteered for this duty. One night my position was overrun by enemy forces. It was not pleasant. I was more than slightly wounded. Do any of you know why I did not die? Dustoff," he continued before we could make a really stupid guess.

"Despite the fact that my unit was still under heavy enemy fire Dustoff came in and got me out along with several of my comrades. When I recovered, I was assigned here."

Silence, golden or otherwise played in the room.

"Fernary-er, are you playing softball up in Phu Bai?" As I shook my head, he turned his attention to Woody and Harry, "How about you two? Any softball players?"

Before Woody could launch into his athletic background Command Sergeant Major Hoadwonick continued, "I therefore assume that all three of you are flying Dustoff missions." Command Sergeant Major Hoadwonick rose from behind his desk and leaned forward, bracing himself on the desktop with his fists.

"I like Dustoff. There are real soldiers in Dustoff. Therefore, I must assume that you, Fernari-er as well as your comrades have become real soldiers. Would that be accurate?"

Our meek nods seemed to satisfy the Sergeant Major.

"I also notice that all three of you have achieved the rank of Specialist Fifth Class. That is almost as good as becoming a real sergeant. Like me. This makes you technically eligible to inhabit an NCO club, my NCO club for example. Do you agree?"

We did, silently.

"That semi-useful corporal you passed on the way in here will take you to our transient barracks where you will stash your gear. Major Durant is due to arrive here tomorrow morning. He will transport you gentlemen back to your unit. In the meantime, you three will meet me tonight at 1800 hours at my NCO club where it will be my pleasure to buy each of you a drink. You will thank me by buying me several. Now unass my office, and Ferrier, I recall what a Terrier is."

That evening I became drunker than I have ever been in my life. Command Sergeant Major Hoadwonick schooled us in the guzzling of "boilermakers", a hellish combination of Seagram's 7 whiskey and draft beer, swallowed in one motion, followed by doing the same thing over and over again. Meanwhile he lectured us on the dangers of marijuana, smoked by "damn hippies", cautioned us on the use of any non-prescription narcotic substances and praised us for being part of "one of the best damn units in his Army." By the time Major Durant rescued us the next morning Woody had thrown up on the helipad, I threw up on Woody and Harry slept like a baby all the way back to Phu Bai.

We were Back In The Saddle Again.

Chapter Sixteen

The Wounded

Private First-Class Richard Martinez of Tempe, Arizona was shot in the chest. The bullet shattered his collarbone, tore through his pectoral muscle and exited his back leaving a gaping hole and a purple cloud of pain. His eyes were wide with panic, his breath came in heaving gulps, his body was rigid and trembling. Flight medic Neal Johnson smoothed a fresh pressure pad on the exit wound, secured the bandage with gauze strips, and placed two sugar wafers onto the soldier's tongue.

"These will help with the pain," Neal whispered before passing on to the next man.

Private Martinez nodded, relaxed, showed a slight smile and closed his eyes. Necco wafers. Candy. I had seen our medics do it before. Bob Byrd and Jack Welsh did it as well. Harry used M&M's. Placebos plus compassion combined to keep the wounded alive until they reached the hospital. The medics knew, they just knew.

Steve Camarano cradled the head of Sergeant Abraham Dunes in his lap. Sergeant Dunes lower jaw was blown away, he was gasping for air and convulsing. Without hesitation Neal drew a surgical scalpel he was not supposed to have and cut a hole in Dunes' throat, made an incision in his windpipe and inserted a breathing tube to keep the airway open. Dunes' breathing eased, the convulsions stopped, and Neal moved on.

"Dave, get over here," Neal ordered through the ship's intercom. I racked my rifle and climbed out of the gun hole being careful not to step on the wounded.

"This is going to bleed like a bastard when I take this off. Take a pad, press down as hard as you can on the wound until I can reset the tourniquet," He said as I knelt beside the litter.

Working swiftly, his hands drenched in blood, Neal removed and retied the tourniquet a bit higher on the shattered leg of Second Lieutenant Michael Patterson. The gush of blood stopped. Lieutenant Patterson stopped bleeding.

Six minutes later we were on the hot pad of the 22nd Surgical Hospital. The three wounded soldiers were off-loaded and rushed to pre-op. Neal, Steve and I hopped back aboard as we lifted off to refuel. This was our third pick up today. It was not yet noon.

Dr. Louis Zoska joined us as we were having coffee in the 22nd Surgical Hospital mess hall later that morning.

"Did you do that traych on that Cav sergeant this morning, Neal?"

Neal Johnson did not look at the surgeon as he mumbled, "Yeah" into his coffee cup. Flight medics were not encouraged to do such procedures which required training far beyond their authorized capabilities.

"It was a nice, clean cut. You probably saved his life by doing it, but you could have just as easily killed him."

"Wasn't nuthin' easy about it Doc," Neal answered looking at the doctor this time.

"Guy was in really rough shape Doc," Steve Camarano added.

"I know that" Doctor Zoska said, "it was good work. I just want you to be careful, you know?"

"We all know," Captain Harmon, the aircraft commander added. "We all know."

Our medics walked a very fine line between what they needed to do and what they were trained to do. Stop the bleeding, clear the air passage, elevate the head were the in-flight priorities, except when they weren't. We saw horrific wounds every day. Bodies torn apart, burned, crushed, bleeding, dying. The medics had to sort them out, prioritize them, bandage, mend and nurture them while flying at one hundred and twenty miles an hour with people shooting at us.

The soldiers treated on our ships were doctored by stressed out teenagers with eight weeks medical training, thrown together, improvised medical kits, authorized and unauthorized abilities and mostly this somehow worked.

"Why don't you come over to the OR later, I'll show you how to do a few things." Doctor Zoska said, patting Neal on the shoulder. "Good work, all of you," he added, rising from the table, "I don't know how you do it."

Dr. Zoska shook his head as he left the mess hall. We finished our coffee and waited for another mission. Which came later that afternoon when we landed on a paddy dike next to where an Armored Personnel Carrier had been destroyed by a command detonated mine sitting atop a fifty-gallon drum of gasoline. The four-man crew had been burned alive, scorched beyond individual recognition. Two soldiers who were riding on top were thrown clear by the explosion, their bodies ablaze. When they landed in the stinking water they rolled and survived. Their burns were extensive, their chances of survival minimal. Both troopers had been mercifully sedated by med-pack morphine before we arrived and still, they moaned in agony as they were loaded aboard our ship. The smell of burnt flesh, gasoline and shitty water filled the ship. Neal motioned for me to join him and began cleaning them up. I was trying not to gag watching Neal peel away layers of burned skin, flushing open sores with clean water, applying burn gel and lidocaine.

"Here," he said, handing me a pile of clean gauze bandages, "after I put on the cream cover the wound with the gauze. Can you do that?"

I nodded. He continued. One of the burned soldiers died, the other we flew directly to the hospital ship Repose where he was rushed below decks to be further treated.

From my gun hole I watched the deck of the USS Repose rise and fall in the swell of the South China Sea. The Repose was one of two hospital ships stationed off the northern coast of I Corps. Repose and Sanctuary, two islands of mercy one mile offshore. Each ship had a patient capacity of around eight hundred patients plus crew. They were always full.

We took the most critically injured to them, head wounds, stomach punctures, so prone to infection, fevers of unknown origin, trauma and shock cases, and burn victims. As we approached the helipad on the rear of the ship the white landing square dipped ten feet or more in the waves then rose up rapidly in the rolling sea. I once asked our crazy ass pilots which was better, to set down at the bottom of the roll and ride it back up or to set down as it peaked and ride it down. They just grinned and said whichever looked like more fun.

We made hundreds of landings on these ships. Never lost a patient, never went in the drink. Whenever possible we would shut down on the flight deck for a few minutes. The crew would unfailingly hustle us out some hot coffee, fresh sandwiches, even, I swear, ice cream.

Once our patients were offloaded, we could sit on the deck, bounce up and down with the waves and look back at the Vietnam shoreline. Even in full daylight it appeared sinister, foreboding, like the deep and dark jungle that hid King Kong. And we lived there.

"Want another one?" Steve Camarano said dangling a ham and cheese, warm bread sandwich in front of me.

"No, thanks, I've had enough." We were still shut down on the deck of the Repose. It was getting on full dark. We had been flying all day and into the night. We were looking toward the blackness of the shoreline where multiple colorful, ominous explosions flashed over and over from the jungle. The charcoal sky turned bright orange, yellow or white as the faint but frequent "karumph" sounds of the shelling drifted across the sea to us.

"Charley must be moving," Steve said, "lots of "HE" out there." "HE" was military talk for high explosive. We saw a lot of that.

"Better there than here," I said, weary to my bones, but relaxed, if that was ever a word one could use in Vietnam.

That was the thing about the hospital ships. A mile offshore and the only place I had been in a year where I was not, or could not be, shot at.

"Where's Pete and Willie?" I asked. They were our pilots, flyboys.

"Flirtin' with the nurses, most likely. Tac push is quiet, we can sit for a while," Steve replied.

Even shut down and scattered about the ship we could be airborne and inbound in two minutes if the radio lit up. It was nice that it didn't.

"How much time you got left, Steve?" I asked.

"Two months, three days," Steve replied.

"Under sixty you ought to get out of the air." I suggested. Unit policy was that when a crew chief got short, sixty days or less, he could get off flight status, work maintenance on the helipad. Some did, some didn't, Steve wouldn't. He grunted a non-answer as Neal Johnson returned to the ship carrying all the shipboard medical supplies he could beg, borrow or transport.

"Lookit' this," he exclaimed, "sterile saline, in plastic bottles! This stuff is great for cleaning wounds." Neal was happy as a child, dumping supplies into the ship. "They got sulfa powder, iodine, ibuprofen, ace bandages, gauze…"

"They have any Necco wafers?" I asked semi-seriously.

"They got Tums, they're like an antacid, same thing only not candy."

Steve shook his head and helped Neal store the loot. Pete and Willie came across the helipad ready to take off.

"We good to go?" Willie asked.

"We're good to stay," Steve replied, all's quiet on the fox mike."

"Anything inbound?" Pete asked.

"Clear." Steve replied.

"Then let's sit a while," Willie said and they turned to rejoin the nurses. "Did you guys get some chow?" He asked. Steve waved his sandwich at him.

"Steve, call home and tell them we're resting the rotors," Pete added.

"Resting the rotors?" Steve asked.

"Technical term, I don't have time to explain it to you, that blonde nurse is going to read my aura." Pete grinned over his shoulder and walked back into the ship with Willie.

We sat on the pad listening to the ocean, watching the onshore explosions, savoring the moment. Not getting shot at.

Chapter Seventeen

Trials And Tribulations

"Not Guilty!"

Sean Patrick Sullivan could hardly believe his ears. The jury foreman had just read the verdict in the kidnapping and rape trial of Kendall Prescott to a stunned and silent courtroom. Margaret Mary buried her face in her hands and wept. Kendall Prescott smirked as his lawyers congratulated themselves. The District Attorney for Arapahoe County slammed his briefcase in disgust. The trial of the "Tuloc Seven" as the case had come to be called in the press had concluded. The defendants, squatters who had built a compound in one of the high meadows, were unanimously acquitted of all charges pending their agreement to leave Colorado, permanently.

"The State of Colorado doesn't like hippies," Corporal Girard explained, "getting rid of them was the only thing the Court was interested in."

Sean was stunned as he led his daughter from the courtroom. The State's case was weak, half-hearted and totally reliant on Margaret Mary's testimony. She was on the witness stand for three grueling days, speaking the truth, which was completely discounted by the perjured statements of ensuing "Tribe" members.

The Court heard testimony that Margaret Mary had come willingly to Tuloc Meadows with the defendant Kendall Prescott and had only been restrained when she presented a danger to herself and other members of the community.

Several members of "the tribe" testified that Margaret Mary was a frequent drug user, a communal leader who frequently overdosed on drugs and was delusional and dangerous before being restrained. Kendall Prescott, they maintained, had been her lover and protector until she became unmanageable. The tribe testified that it was their intention to take Margaret Mary to a medical facility as soon as the winter snows allowed them to leave the meadow and that they were prevented from doing so only when Corporal Girard illegally searched their compound.

Despite his contradictory testimony the jury and the court found in favor of "The Tribe", banishing them from Colorado but exonerating them of any wrongdoing in their treatment of Margaret Mary.

Margaret Mary and Sean returned to Chicago in shame and disgrace. Corporal Girard vowed to make it his personal business to see that all of the members of the "tribe", and particularly Kendall Prescott, were out of Colorado within twenty-four hours.

"If God is a merciful God and a just God, how could he have allowed this innocent woman to have suffered so much and her tormentors go free?"

Peter was more than perplexed. Anger raged within him and shook his emerging beliefs in the Catholic philosophy.

"God is a just God, a merciful God, men however are neither just nor merciful, not all the time. This was the judgment of men, not God." Father Wyman, Peter's spiritual advisor said.

"It's wrong! Wrong to the core!" Peter raged.

"Men are often wrong, that is why we need God," Father Wyman advised. "I am going to speak with you frankly Peter, you may not like what I am going to say but think about this in light of the vocation you have chosen."

Peter agreed, Father Wyman continued, "Your friend Margaret Mary chose to go live in a commune with a man she was not married to but was sleeping with. Correct?"

Peter nodded, irritably.

"In this commune drugs were used, and your friend also chose to participate, however sparingly, in their use. Correct?"

Peter didn't nod this time. He was wavering between indignation and a fierce desire to defend Margaret Mary. With considerable effort he remained silent and attentive.

"Perhaps if she had not made those choices, she would not have found herself in a situation resulting in abuse and debasement. Correct?"

Peter was reaching his tolerance limit for the word correct, but he remained silent.

"God's rules may seem antiquated to some, rigid and unreasonable to others, foolish and unsophisticated to many, but they are God's rules. They are intended to guide us through this life with a set of moral guideposts to mark our way. People tend to question those rules, ignore those guideposts and behave in ways that are against the teachings of God, the teachings of this church. When they do that the consequences may be part of God's judgment."

"But what about the ones who testified under oath to their lies?" Peter finally exploded.

"Perhaps their consequences will be at another time, in another way. We must believe in the righteousness of God, the way he manifests it, not the way we want it to happen. That is an essential part of the faith you will need to pursue this vocation."

Peter sputtered, then reflected, allowing his mentor's words and reasoning to sink in. What he said sounded right, but this was Margaret Mary! She did not deserve this!

Peter had much to learn about what men deserved and what men received.

"Self blame and self pity are two diversions an intelligent young lady like yourself should not allow herself to indulge in for very long."

Margaret Mary listened carefully as her counselor at the Trauma Treatment Center, Lois China Adams, repeated the admonition gently and patiently.

Margaret Mary had been seeing Lois for six weeks. They were working together through the grief and guilt and sorrow of Margaret Mary's experience with Kendall Prescott, with compassion leavened by reason.

"You were mistaken in your affection towards Kendall. You were misled by a skilled sociopath whose dysfunction greatly outmatched your common sense and maturity. Once you accept this, you accept your humanity. Humans make mistakes, they are built into our hearts and our minds. We must learn from them and move forward, not wallow in them and learn nothing at all."

"It hurts," Margaret Mary whispered.

"I know it does dear. The only way I have ever found to lessen that hurt is through self forgiveness. This is what we are here to do."

Margaret Mary took all this in while stubbornly clinging to self-recrimination and bitterness toward Kendall and his followers.

"You know something of my background do you not? Lois asked.

"There are lots of stories about you. I don't know which ones are true."

"My father sold me into prostitution for six hundred dollars when I was twelve years old. He delivered me to a brothel just off Canal Street in New Orleans."

Margaret Mary's eyes were wide with shock and wonder.

"That must have been horrible, " was all she could think of to say.

"Yes, yes it was, but not all of the time. That brothel became the first safe place I had ever lived in. I was a precious commodity, what was called a "premium visit" in the house. The other girls pampered and protected me. Men were not allowed to cuss or swear in my presence and serious misfortune would befall any customer who mistreated me."

"But still," Margaret Mary began.

"Yes, but still." Lois went on, "I remained in that house for five years, until one of the girls burned it down. As we all ran out of the flames, she ran into them."

Margaret Mary's gasp elicited a sad smile of remembrance from Lois.

"I never tried to run away from that house before the fire. Not once. I felt safe there, I was cared for there. After the fire I was taken in by a local family. They treated me well. I ran away from their home several times. I wanted my old life back. I was seventeen and very foolish."

"What did you do?" Margaret Mary asked.

"I persevered. I lived on, educated myself and forgave that little girl for liking her time in that house. I stopped running away and even managed to forgive my father for selling me."

"You forgave your father?" Margaret Mary could not disguise her amazement.

"Only after I forgave myself and had to let go of the anger I felt for him."

"What happened to him?"

"He took the six hundred dollars he got for me and drank himself to death less than a month after he sold me. I didn't hear of this for many years, when I did, I felt sad, for him and for me."

"I don't think I'm ready to forgive Kendall," Margaret Mary asserted.

"Then let's start with you and see what happens after that."

"You gotta' keep that left hand up, lieutenant. You don't want to be stopping every punch with your nose."

Teddy was loving his assignment at Quantico. Getting paid to go to the gym every day, plus three hots and a cot, all the coffee he could drink and the satisfaction of serving his country. Life was good, at least on post, and always with the Corps. Off post, on the city streets was another situation entirely. Anti-war protests and demonstrations were an everyday event. Service men and women were often targeted by protestors, harassed and mocked as they went about their daily lives. To a large portion of the American public serving your country in the military was a thing to be scorned. Teddy was not wearing this well.

"Sergeant Gianoulos, you are accused of assaulting a civilian in a public tavern. Is that correct?" Captain Jenkins, Teddy's commanding officer asked.

"Yes sir."

"Did you in fact assault a civilian in a public tavern, Sergeant?"

"Yes sir."

"May I ask why you assaulted this civilian?"

"He tried to rip my service bar off my uniform sir," Teddy replied, suppressing any emotion.

"Your service bar?'

"Yes sir, along with my Bronze Star ribbon with Oak Leaf cluster, Combat Action Ribbon and two awards of the Purple Heart, sir."

"I am aware of your resume Sergeant Gianoulos. What prompted this action by the civilian?"

"I believe it was an inappropriate amount of alcohol sir."

"And were you drinking as well Sergeant?"

"Moderately, sir."

"How many times did you strike this civilian, Sergeant?"

"Twice sir. Left, right combo."

"And that brought an end to the altercation?"
"It did for him sir."

Captain Jenkins tried to suppress a smile though he did not order Teddy to stand down from his position of attention.

"Sergeant Gianoulos, we are in a somewhat precarious situation here. We must interact with the civilian population in a positive manner when we can. It appears, and the statement from the bartender backs you up, that you were unquestionably provoked by this individual. In my experience, however no Marine should require two punches to handle such a situation. You may consider this a severe reprimand, which will not appear on your records, and an admonition to either strengthen your left jab or skip it altogether and go to the crossover right. That is all Sergeant, you are dismissed."

Teddy exited his CO's office and headed directly for the NCO club. Happy Hour was ticking away, and Teddy did not intend to miss more than a minute of it.

"In the Old Corps a Marine walked down the street all the pogey-bait in town got the hell out of his way." Gunnery Sergeant Francis Kessler was holding court as he did every afternoon from five to seven PM while downing jumbo steins of beer in one swilling swallow.

"A Marine in uniform couldn't buy a drink in this town in the old days, respect, that's what it was all about, respect." Gunny Kessler had been in the Corps since, "Christ was a choir boy." Teddy loved listening to his stories, loved even more being part of his brotherhood.

"Gianoulos", Kessker roared, "I heard it took you two punches to knock out a hippy the other day. What kind of sissy-ass Marine are you son?"

"The kind of Marine who owes you a beer, make that two beers, Gunny, one for each weak-ass punch," Teddy responded bellying up to the bar.

"Hard to argue with that kind of logic," Gunny Kessler replied, "Sammy, give this candy ass a double shot of Jack Daniels, one for each weak-ass punch."

Teddy was learning in Quantico what Dave was learning in Phu Bai, hanging around a bar with the Old Guard non-coms was very hard on the liver.

Chapter Eighteen

Letters To And From

April 12, 1969

My Dearest David,

Please don't be upset with me because I have taken so long to write you. When Peter told me you were in Vietnam, with Teddy, I was angry, and frightened, and distressed that you had placed yourself in such danger. I was frightened and worried for Teddy too, but I never pictured you being there, I thought you were safe, in school. Like I wasn't.

I made such a fool of myself with Kendall. That is all over now, I am back home with Sean and my Aunt Rose. I am getting better every day and learning to forgive myself for my foolishness. This is very difficult for me and there is much to tell, so get home soon, I need to see you, to talk to you to make myself better.

It is too hard to write about my troubles when you are in that terrible place. Please, please, please be careful. Peter told me about what you are doing, rescuing soldiers and taking them to hospitals, I should be proud of you, and I am, but I am so frightened you will be hurt. Come home to me please, I want so much for all of us to be as we were. Remember that night in the Pewter Pot when we promised to meet back there if anyone ever tried to make us something we didn't want to be? I need us to keep that promise. Come home safe, you must.

I started to count the days since I last saw you and when I reached 500 I couldn't help myself, I started to cry. I hate acting like a silly girl but when I remember that I saw you every single day (almost) since the fourth grade and now we are so far apart and for so long my heart just aches and aches.

I must stop because I want this to be a happy letter. I am back in school and enjoying my classes. I am talking to a wonderful counselor, and she has helped me so much. Please write to me and tell me all the kind lies about how safe you are and how well you are doing. I won't believe you but I will take comfort in your saying so. Sean sends his blessing and a prayer. I send both as well, along with all my love,
Write to me soon,   MM

****

April 20, 1969

Hi Mags,

It was wonderful to hear from you after all this time. I have been worried about you as well. When we talked last Christmas, I was hurt and angry that you wouldn't let me come see you but I now understand you needed to get better your way in your time. Everybody's been a lot vague about what happened between you and Kendall and I understand that too. We can talk when I get home. With a little luck, I've only got five months, two weeks and three days left in the Army and by the time you get this letter that will be five months, one week and five days, but who's counting?

I'm glad you are back in school. I don't believe I have ever met, or ever will meet a person less in need of forgiveness than you. Whatever bad happened between you and Kendall unquestionably came from him, not you. I can say that and believe it without knowing a single detail. You see, I know you and there is no bad in you. Kendall will be another topic for another time.

Now for the sweet lies, most of which are true. I no longer fly rescue missions. My commanding officer took me off flight status two weeks ago. He said enough was enough and made me a radio operator. This is true, I swear it.

I am in a place called Phu Bai in Northern I Corps. When we got here Phu Bai was the wild west, lots of shootin' and scootin'. We are now a major firebase, thousands of guys, with an airstrip, artillery, perimeter security, and lots of dead foliage outside our wire. Really, we are pretty safe here. This is still Vietnam and I will not lie to you, anything can happen, but what does happen, happens a lot less and

I know my way around pretty good. I will not take chances, foolish or otherwise. I want to come home in one piece. I want to see you and tell you all I can about this. I'm going to need some getting better as well. We can help each other, okay? I've got to go now, I've got a radio shift, hours of tedium, moments of high drama. Please write soon, I am so glad to know you are well,

Your confused about our relationship but still very best friend who loves you, DF

****

May 3, 1969

Dear David,

When your letter came today I couldn't wait to open it. I read it outside, at the mailbox while it was raining. I got all wet, your letter didn't. I am so glad to hear you are in a safe place, safe for Vietnam anyway, and that you are not flying in helicopters anymore. Please, please stay as safe as you can and come home to me, alright? I looked up Phu Bai on the map. You are so close to North Vietnam, are you sure you are as safe as you are telling me? The TV and the newspapers have stories every day about the increase in fighting, the terrible battles, the number of our soldiers being killed. This is very frightening to read about, I cannot imagine how you endure all that. I want to know so much about what you are doing and where you are but if you do not want to write about it now that's alright too. I understand. Some of my questions may sound silly, but I think about you a lot and want to know. Do you sleep in a building or in a tent or outside on the ground? Do you get enough to eat? What do you eat? Is the food good? Can you send a picture. I would love to have a picture, some pictures, lots of pictures that show me you are safe. Please.

The news here is always horrible about the war. Many, many people are protesting, wanting us out of Vietnam and bringing our soldiers home. Me too. I can't make any sense out of what the politicians and newscasters are telling us. If we are winning, why haven't we won? Why are so many of our soldiers being killed and wounded? When and how will it all end? I know these are questions you probably don't have time to ask yourself and I must believe that if our country is there, if you are there, we are fighting for something good. I'm just not sure of what that good is. Please tell me your thoughts.

Do you have any books to read? Do you ever have time to read them? Are you doing any writing? You really should, what you are doing deserves to be remembered. My Aunt Rose can't wait to meet you and Sean speaks of you daily. You are in our thoughts and prayers every day. Please write soon,

Love, MM

****

July 7, 1969

Hi Mags, First of all, all is well here. We are not involved in any major operations and enemy activity in this area has been very quiet since the end of the A Shau Valley campaign. Truth. You are probably looking at the pictures. The little girl between me and my very good buddy, Woody, is an evacuee from a nearby village that got burned out by the Viet Cong. She was here with us for several days before being relocated to a more secure refugee area. It is amazing, absolutely amazing, how positive and happy some of these people can remain in terrible circumstances. This little girl was friendly, smiling, happy and grateful in a world where little of that should be possible.

The rest of the pictures should give you some idea of how we live. We now have barracks, called "hooches" which got us out of the tents we lived in for too many months. We have electricity from a noisy generator which breaks down often, running, or should I say falling, water from a steel tank set atop a tall platform and a hospital mess hall which serves decent, hot food. Primitive as all this may seem, we live better than the troops in the field who sleep on the ground, no showers, C-ration food and constant deadly danger.

We don't get much news about politics and current events. The newspaper here is called The Stars and Stripes and pretty much filters what we know. We can receive magazines and newspapers from back home but they seem unreal to us, to me anyway. I have learned to live in the right now, where I am, what I am doing world. This keeps me and my comrades safe, or safer, without the distractions. The larger questions can be pondered after, after I get home. Which is in about three months. I'm not sure of the exact date yet, I am trying for an early discharge and it's a little complicated. If all goes well I should be back in early October. Every day that passes is one less on my tour and I must tell you that the closer I get to coming home the more anxious I become about my surroundings. Yes, I am being careful. Very careful. I read a lot here. There is little else in terms of distraction. Mostly I get pulp paperbacks from the PX but writers like John D. McDonald who writes the Travis McGee series and Donald Hamilton who writes about Matt Helm are very good and underappreciated. I read a terrific book, "In Cold Blood" by Truman Capote whom I believe also wrote Breakfast at Tiffany's. First rate. I'm going to let the heavy thinking wait till I get home, meanwhile I keep busy, stay out of harm's way and miss you. Write Soon, DF

****

July 18, 1969

Dear David,

Thank you so much for the pictures! That little girl looks delightful. Please tell me she is safe. I can't get over how you and your friends look. You are smiling, they are too, you must be doing wonderful work to shine so. I have read "In Cold Blood" and think it is very well written but such a sad story. I am going to look for

Travis McGee this week, Sean says he can't be all that bad, Irishman that he is. One of my summer courses is Psychological Diagnosis and Treatment of Trauma. Nobody will wonder why I took that one, but what I am thinking about in the class is you, what you have gone through and seen there and how it might affect you later. I know you have done nothing but good but so much that is not surrounds you. We will talk about this for hours and hours when you get home, okay?

I spoke with Peter on Tuesday, there is a chance he can go to Rome to continue his studies at the Vatican! Imagine. What different paths we have all taken. I must choose a Major for my degree soon, so far, I have taken general interest courses while I sort out my options. I thought of teaching of course but I cannot convince myself I want to do that for the rest of my life. I am interested in certain kinds of social work, disadvantaged children and single mothers for example, but that too may not be a life course I want to pursue. Right now, today, I do not have to make that decision so there's another thing you and I must talk and talk and talk about. The other day I was remembering the wonderful time we had in York Beach so long ago. Do you think we could go there when you get home? Just for a day or two, three even, just us. Think about it, please. Sean is reading his poetry at the library tonight so I must get ready. If you have any more pictures, I would love to see them. I will save them for you if you cannot get copies. Love from all of us, especially me. MM

****

August 14, 1969

Hi Mags,

I'm sorry for the delay in writing back to you but I have been down in Da Nang where a really offbeat, but friendly Sergeant Major, who used to hate me, is trying to teach me to drink myself to death. Just kidding of course, but Oh My God, can that guy drink! He loves Dustoff soldiers and has become my friend in a very alcoholically way. How's that for a word? "Alcoholically," I like it.

Severe hangover aside, the trip was a sort of in-country R&R, that means "Rest & Relaxation in the military world. I got neither but I appreciated the gesture. The days are counting down now, I am way under 100, officially "short" which means I have a short time left. I have been taking more pictures, I must keep my memories of this place sharp and clear, most of them have been good. I have witnessed incredible acts of sacrifice and heroism almost daily. The men I am here with are heroes and it would be a tragedy if no one ever knew that. Something for me to write about, someday. The pictures enclosed I begged from other guys; I have a lot of undeveloped film I will be bringing home with me. As hard as some of this has been I don't believe I will ever regret my decision to join the Army or to question the amazing good fortune

I had in being assigned to a medical evacuation unit. These men are the best of the best and I am honored to be among them, always will be, I hope. Before we get off the subject of pictures would you send me one of yourself? I told Woody the other day that you were the prettiest colleen on either side of Dublin. He said he didn't know what a colleen was, and I want to show him. Thanks in advance because he'll know when he sees your picture,

Lots of love DF

Margaret Mary and I exchanged several more letters. Her picture arrived and Woody discovered what a beautiful colleen should look like.

Then one night as I lay dreaming, no doubt of how few days I had left in Vietnam, a 122 mm North Vietnamese rocket slammed into the swamp just outside our hooch. Fortunately for us the soft, mushy, muddy, disgusting swamp water absorbed much of the resulting shrapnel, but the blast itself, the concussion from the explosion, blew the back door and window screens off our hooch, lifted the edges of the tin roof straight up and scared ten years off my life and the rest of the guys in the hooch. Miraculously, once again, no one was seriously injured. However, our belongings, personal gear and bedding were blown all over the unit area. By the time we peeked out of the bunker and assessed the damage one of the guys found my letter box, the one holding letters received, letters saved, and mementos cherished submerged in a foot of that crappy swamp water. Not all of the correspondence was lost but the ensuing letters, those I received after Margaret Mary's picture, were reduced to mush.

"C'est la guerre."

Chapter Nineteen

DEROS

Then, not so suddenly, my day arrived. October 6, 1969. Date Estimated Return From Overseas, End of Tour, end of enlistment. Walkout day. I was going home. Alive. I made it.

I still had to get from Phu Bai to Da Nang, Da Nang to Cam Rahn Bay and Cam Rahn Bay to Seattle and from there to Boston and then home. Sounds like a lot, it wasn't, I'd already made the trip a thousand times in my head. Piece of cake.

Man plans, God laughs.

My very good buddy, hooch-mate, comrade, Danny Hernandez, "Danny H" was leaving with me. Don Avila and Woody rotated home three weeks earlier. John Kane as well. Harry Giles was, as you can imagine, another story. Not a good one to tell. Maybe later.

The old Fort Meade/Phu Bai gang was breaking up. Fresh New Guys replaced Frosty Old Guys. I didn't bother to get to know the New Guys, they were less to worry about after I was gone.

New fly boys arrived in our unit fresh out of Fort Rucker, Alabama flight school. They all looked about twelve years old. Seasoned pilots Ken Harman, Pete Lewis, Willie James and the like retrained them with patience and tolerance. Their cockiness was soon replaced with caution and the experienced pilots rotated home as well. Crazy way to run a war.

On my last morning in Phu Bai, I looked at the pile of junk on my bunk, thought of how vital this stuff was yesterday, how irrelevant it was today. I already turned in my pistol and M-16, but I still had my PX cowboy holster and ammo pouches. I gave these to one of the new guys, a patient protector I had half trained who was still putting together his flight kit. I left the once precious poncho liner on my bunk, a lightweight, nylon blanket, ideal for the tropics, too bulky to bring home. The rest of my flight gear, fatigue uniforms, towels, boots, flip flops, paperback books and sundries stayed as well. I packed one change of uniform, my undeveloped film, shaving kit, and the latest Matt Helm paperback and walked out of what was no longer my hooch.

"Godspeed, Specialist," Master Sergeant Bunderson said, shaking my hand and handing me my travel orders, 201 file, finance records and medical folder.

"Sergeant Hernandez, it has been an honor and a privilege. Say hello to that beautiful wife of yours for me. Travel safe." Another handshake, another treasured moment.

The New Guy company clerk looked up from his typewriter and nodded as we left.

Waiting for us on the helipad was the rest of the unit. The Fresh New Guys, the semi-fresh New Guys, even a few staff and nurses from the hospital. The farewell tradition was the ship would make two low, very low, high-speed passes over the pad as everybody waved goodbye. Colored smoke grenades were taped to the skids and popped during the flyby. After the second pass the ship would bank left, head south and start us on our journey home.

After a heartfelt chorus of "Take Care's", "Be Safe" and even a Sayonara or two we climbed aboard the ship, lifted off, soared over and above the helipad twice and headed south. The fondest and best goodbye I ever had.

I didn't exactly know where the great lump in my throat came from, or why I had tears in my eyes, but I did. It would take many years for me to understand why.

"Get in here, Ferrara-ay," Command Sergeant Major Hoadwonick yelled from inside his office. In order to clear post and continue on our homeward journey Danny and I had to clear personnel in Da Nang, where Command Sergeant Major Hoadwonick was waiting for us.

The first thing I noticed was that I was apparently back to square one on my name pronunciation. In spite of surviving Command Sergeant Major Hoadwonick's frequent visits to our unit and many, many, bleary happy hours and hours and hours of trying not to die of alcohol poisoning while the Sergeant Major taught us how to drink "like real soldiers", I was back to being name mangled once again.

I entered Command Sergeant Major Hoadwonick's office and locked myself in at attention in front of his desk.

He looked up at me, then down at a sheaf of papers, then rocked back in his chair.

"Furnier, I am officially inviting you to stay in my Army. Despite your inauspicious start you have managed to become worthy of remaining in the Armed Forces of the United States of America. All you have to do is sign here and I will see that you are afforded this privilege."

"But Sergeant Major," I stammered, knowing I was walking on shaky ground, "I don't want to continue on in the military. I want to go home."

"The Army is your home Ferrera-ay, haven't you learned that?"

"Yes, Sergeant Major, for the past three years the Army has been my home, and it has been a privilege to serve in it, but I have good reasons for not planning to stay."

Command Sergeant Major Hoadwonick huffed and puffed and shook the sheaf of papers at me. "Give me one good reason, Fernerier, why you should not stay in my Army."

I realized I was at a crucial moment. I did, however, have the perfect answer.

"May I show you something, Sergeant Major?"

"It better be one hell of a something Furrierer.".

From my shirt pocket I took the carefully wrapped picture Margaret Mary sent me and handed it to Sergeant Hoadwonick. He looked at it for a long moment, then smiled,

"That's the prettiest colleen I've seen this side of County Clare."

"You know the Irish, Sergeant Major?" I asked, respectfully.

"The Hoadwonicks are descended from a long line of Irish Kings, Fyrrerrier. Didn't you know that?"

"I'm learning more all the time, Sergeant Major."

Command Sergeant Major Hoadwonick rose from behind his desk and held out his hand for me to shake. I did, then, as I did many years before at Fort Jackson, South Carolina, I got the hell out of his office.

"How are we doing, Danny?"

"Admirably," he replied, "I would say we are doing admirably."

Danny and I were sharing a two-row seat on a jet airliner somewhere over the Pacific, west of Japan and closer and closer to Fort Lewis.

Danny was drinking a Heinekin beer, and I was sipping a Crown Royal with ice. We hadn't slept in two days, due to excitement mixed with processing, and were getting the slightest bit dopey as the miles rolled under us.

"What's the first thing you want to do when you get home?" I asked blearily.

"Haven't I shown you a picture of my wife?".

Oh, yeah, wife. Didn't have one, but I got the drift. "How about after that?" I persisted.

Danny took a deep breath, then a deep swallow of his beer. "Deep dish pizza and lots of TV."

Yeah, I could do that. The more I thought ahead to what was coming, what I would do, what I could say, the more anxious I became. I yearned for Margaret Mary, far away in Chicago and wounded, remembered small town Lowell, gray and drab, thought of my family, them the same, me really different, and knew deep dish pizza and TV were not going to cure my ills. But I was glad for Danny, ills and all.

"You know what else I really want to do?" Danny added, "I want to go for a long ride in my car, with Brenda, with the windows down, radio blasting, going nowhere, just cruising."

"With nobody shooting at you," I added, throwing a turd in the punch bowl.

"Yeah, that too," Danny said dolefully, finishing his beer and waving for another.

Our plane flew on as I dreamed of deep-dish pizzas, long car rides and guiltily, Danny's beautiful wife, Brenda. There was a steak dinner when we landed at Fort Lewis. It was the middle of the night, but the mess hall was brightly lit and we were served our dinners at the table by wide eyed, freshly minted buck privates performing KP while completing basic training. Nice.

What followed was a succession of long lines, long forms, a physical exam and a fitting for a dress uniform, replete with all our authorized awards, decorations, rank and ribbons. Nice.

Danny and I went our separate ways at one point in the process. He was going on to another duty station, I was being discharged. We made a plan to meet up when we were finished and travel together as far as we could. He went into a building marked Permanent Change of Station. I went into a building marked Expiration Term Of Service. I persevered through the poking, prodding and processing through the night. The Army was apparently trying to do us a solid by getting us through the system and back to our homes as quickly as possible.

Then one of the most melancholy and troubling incidents I experienced during my about to end military career happened in a barracks latrine, shortly after we had been issued our beautiful, beribboned, dress green uniforms. I had never been issued dress greens before. Summer tan khakis and olive drab fatigues, jungle, Nomex and otherwise made up my military wardrobe so far. All of the men I was with, like myself, were dressed in the jungle suit we left Vietnam in two or three or four days ago. Between out-processing, day long plane rides, in-processing and out processing again we had neither slept nor showered for too long a time. We were not at our most dapper. Until our new uniforms were issued.

We carried them reverently from the post tailor's office to a barracks where we could shave, shower and put them on. They were on a hanger and covered in a thin plastic bag, like a dry cleaner bag. Through the filmy wrap we could see brass buttons, gold stripes and rows of ribbons. Nice.

Along with the scruffy crew I travelled with for the past few days, we stripped down, shucked our weathered fatigues, and hurried through a steamy shower and quick shave. Then wrapped in towels or fresh new boxer shorts we tore open the plastic bags covering our dress uniforms.

I had never seen one of these with my name on it before. My E-5 rank, never important in Nam, was emblazoned on each sleeve. Polished brass lapel insignia with U.S. on one side and a shining medical caduceus on the other, a row of decorations, National Defense, Vietnam Service and Campaign, Air Medal with tiny oak leaves for multiple awards, a silver Aircraft Crewman's Badge, even a Good Conduct Medal. I was moved, proud, happy, and when I turned around and saw the same look on the men around me as they contemplated their uniforms. I was humbled.

That kind of goofy looking guy who sat across from me on the plane, Buddy Holly glasses, buck teeth and a scruffy, dusty set of faded jungle fatigues was tenderly touching his jacket which held a Silver and Blue Combat Infantryman's Badge, Purple Heart, Bronze Star and Army Commendation Medal atop the standard service bar. He was an E-5 Buck Sergeant with eighteen months of overseas service bars sewn on the sleeve of his uniform. Like me he had never seen or imagined himself in such a uniform. It was covered with honor.

Throughout the barracks it was much the same. Ordinary guys, rumpled, ragged and shaggy one hour ago were trying on brand new uniforms with their name, rank and service badges proudly displayed. I was not the only one in the room feeling very emotional as I saw and recognized what kind of company I was in, and part of. Soldiers, veterans, patriots.

"You a sergeant?" The playful banter echoed through the barracks, "I know you six months never knew you was no sergeant!" One soldier laughed as he checked out a buddy's uniform.

"You wearin' jump wings? You mean to tell me you were fool enough to jump out of a perfectly good airplane?" Another joked as his friend pulled on his uniform coat.

"Purple Heart ain't shit! I got three Purple Hearts, that's why they sendin' me home."

And on it went around the room as I went into the latrine to find a mirror and knot my tie. As I pulled through my old faithful double Windsor, I noticed a soldier watching me. It was the guy who had three Purple Hearts. He asked shyly,

"Would you tie my tie? I don't know how."

And I did, then as I did, another soldier approached me holding out his tie, then a third soldier. I tied four more after that feeling sadder and sadder each time. The Army could teach these guys how to field strip and fire a machine gun, load a mortar tube, hump and repair a field radio, clean and mend a gunshot wound, but could not be troubled to teach these men how to knot a tie? They were trained to kill, they learned to survive, they got out of Vietnam alive and weren't taught how to wear their own uniforms? As much as I wondered what they did know, how they did survive, what now useless military skills they were master of, I wondered how ill prepared they were, I was, for what lay before us.

I found Danny as he came out of a barracks next to ours. He was dressed in a plain, ill-fitting, rumpled set of khakis, no ribbons, no rank, not even a name tag. He still had on his funky, Phu Bai jungle boots. He looked embarrassed; I looked great.

"Wow!" he exclaimed, "Looking good! How did you rate the outfit?"

"It's my going home suit, Danny. They just told us we're having a discharge ceremony at the post theatre at 4 PM. It's going to be on TV!"

"How about you? Why the shabby duds?"

"They told me I was supposed to bring my own uniform. I told them I didn't have one. They said I should have talked to my supply sergeant. I told them I was the supply sergeant!"

"This was the best they could round up?"

"It is if I want to make my 2 PM flight to Chicago. How about you? You ready to go?"

This was not good news.

"Dan, they just told us we have to go to a discharge ceremony at 4 PM. It's going to be on TV. My flight's not till 7 PM."

"Hey, I can wait, catch a later flight."

"No way, Dan. Get yourself home."

So, we parted with a handshake and an embrace, military style. Two friends saying goodbye who had been together, every day, every minute for the past year. Life goes on, but I would see Dan again. Another story for another time.

"When you gentlemen approach the podium take your discharge certificate in your left hand and shake hands with the Senator with your right. Try not to stare at the cameras as you approach. Any questions?"

We were seated in a large auditorium rehearsing the ceremony which would officially discharge us from active duty. There was a stage, several flags, lots of lights and cameramen who were milling about, selecting angles.

The bird colonel who was addressing us was being painfully polite, never a good sign in the military. Colonels, bird or otherwise, told, they did not ask.

A hand shot up. "When we get that there certificate, are we out of the Army?" A voice with a twang from deep in the South asked.

"Officially you are still in the Army for forty-eight hours after your discharge. This means that you are still subject to military regulations and protocol until you reach your homes."

"We can go to the airport when we're done here?" Another hand, another question.

"Buses will be waiting in front of the transient barracks in one hour to take you to the airport. If that is all gentlemen, we can begin."

A white haired, gray suited, polished looking guy came in surrounded by assistants. He was Senator somebody or other and would hand out our discharges. TV lights flashed on, a public address system squeaked, and the ceremony began.

"Abbott, James, T." the senator announced without introduction or fanfare. We had been seated alphabetically and Abbot, James T. rose from his chair and approached the stage. The senator presented him with his discharge, shook his hand and turned back to the podium. An assistant waved Abbott, James T. off the stage.

"Allen, Michael, W." The litany continued, certificate, handshake, exit. I was three rows back, leaning forward in my chair, waiting almost patiently for my name.

"Christopher, Alan, M." but after the exchange the TV lights suddenly went off, the cameramen started packing their gear and the senator waved at the rest of us and left the stage. The bird colonel then took to the podium.

"That is all gentlemen, the rest of you may pick up your discharge at the rear of the auditorium."

It was like the air went out of the Good Times balloon. We looked at each other, stunned, but not really surprised. We were, after all, still in the Army. At the back of the auditorium a Lieutenant awaited us. There was a large stack of brown envelopes piled on a table. They contained our discharge papers, travel orders and an invisible handshake from Senator somebody or other.

"There is one more thing, men," The lieutenant announced, "There has been some trouble at the airports, hostility directed at uniformed soldiers in transit. You are thereby authorized to travel in civilian clothes if you so desire."

I didn't have any civilian clothes. I was momentarily and temporarily proud to be wearing this beautiful uniform. With that the lieutenant turned and left. We sifted through the piles of envelopes and followed him outside.

Welcome Home, GI. The worst is yet to come.

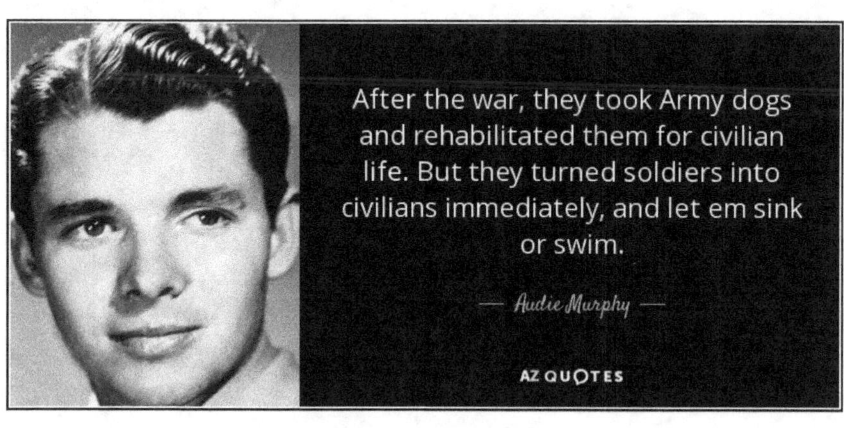

Chapter Twenty

Greetings and Goodbyes

The skyline of Boston slowly materialized out of the early morning haze. I had a window seat on United Airlines Flight 724, non-stop from Seattle. I had been in the air six and half hours. It was a little before 7 AM, October 8, 1969.

The cabin lights winked on, dinging things started dinging, sleepy passengers woke and stirred in their seats. The questions, "What do I do?", "What do I say?", "How do I act?" played over and over in my head. My hands were shaking and if I could have stood up I'm sure my knees would have been weak. Two years, seven months and two days since I began my journey. Twenty-two months and twenty days of that in Vietnam, now it was ending.

As the plane braked to a halt at our gate a stewardess, one I spoke with briefly during the flight, performed an act of great kindness which I have never forgotten. Before opening the cabin doors, she announced over the intercom, "Would you please keep your seats for just a moment longer. Thank you. We have a soldier aboard coming home from Vietnam. His family is waiting in the terminal. Let's let him exit first."

Heads spun around looking for…me. The stewardess motioned for me to come forward. I rose and walked stiff-legged and wobbly toward the front of the plane. The passengers looked at me with curiosity, faint smiles and even some respect. As I reached the cabin door the stewardess applauded, and the clapping spread throughout the plane. I stopped in the doorway, red-faced and grateful, waved to the passengers and the stewardess and stepped into my new life.

My Father and brothers were waiting for me. I noticed my Mom had not made the trip. No one holds a grudge better than my Mom. I had barely heard a word from her during my last ten months in Vietnam. A line scribbled at the bottom of my Dad's letter, placed there, no doubt after some debate. My Mother was furious with me when I extended my tour and went back to Vietnam voluntarily. It showed.

My brother Bob sported a mustache. John's hair was stylishly well over his ears and down his neck. They each wore bell bottom trousers, a fashion I had vaguely heard of but never seen. Everyone was smiling. Then I noticed Teddy, in full Marine regalia just behind them. He had the biggest smile of all. Semper Fi.

My Dad hugged me hard when I entered the terminal. My brothers hung back but we hugged as well. Teddy waited. He held out his hand for me to shake. I hugged him like a bear.

"Welcome home, Dave. Glad you made it," He exclaimed when I finally turned him loose.

"I'm glad I made it too. Thanks for coming."

Sometimes good things repeat themselves.

"Luggage?" My Dad asked.

"Just what I'm carrying," a blue gym bag with film, a few trinkets and lots of memories.

"Where's Mom?" I asked, getting the question out of the way.

"She's home making a big breakfast for all of us," Dad lied. I felt bad for him.

The ride home was quiet, semi-comfortable, a little foreboding and over too soon. Teddy said goodbye in our driveway promising to be back after I got some rest. He said he had a surprise for me. I was dreading the surprise in the house.

My Mother came out of the kitchen, drying her hands on a dish towel. She kind of glared at me, then shuddered, and held out her arms. We hugged hard. She sniffled, recovered, straightened up and asked, "Are you sure you're home for good this time?" It wasn't a question; it was a command.

I assured her I was, much to the relief of her, myself, and my father and brothers who had stayed out of range of the potential fireworks.

Truce be told, my family reunited around the kitchen table and enjoyed a French toast, sausagagey, peaceful, breakfast. Remembering the trinkets in my bag I fetched them to the gathering. For my brothers I purchased rings, a black star sapphire for Bob, a blue gemstone for John. For my Mom there were teardrop jade earrings from Thailand and finally a pearl faced, 24 jewel, self-winding, top of the line SEIKO watch for my Dad. Eyes got wide, appropriate thank you's were received and my eyelids began to slam shut.

"Guys, I don't remember the last time I slept. I think its been two days, maybe three. I've got to get some rest, okay?" I half spoke, half mumbled as the travel, excitement, anticipation and relief fatigue washed over me. Though it was barely ten in the morning I took myself to bed. My old room, single bed, clean sheets, familiar bedspread, cool pillow, oblivion.

The next thing I heard was a gentle knocking on the bedroom door. I tried to ignore it; it wouldn't go away. Then my brother Bob cracked the door, "Teddy's here. He says he's got a surprise for you."

"What time is it?" I managed to mumble.

"Quarter to five, you slept all day."

It didn't feel like it as I swung myself out of bed. Teddy and surprise being my motivation. Whatever his surprise was, a bigger one awaited me when I slid open my bedroom closet. Empty, bare, a few dangling wire hangers, no clothes.

"Mom gave all your clothes away after Ricky Burke got killed," Bob said from the doorway.

"Ricky Burke got killed?" I stammered.

"Yeah, in February. He got drafted."

"Killed in Nam?" I knew nothing of this.

"Yeah, Mom flipped out. She wouldn't stop crying. Then she packed up all your stuff and gave it to the church. She said you probably wouldn't be coming home either."

Bob sounded embarrassed. I was dumbfounded. Teddy was waiting. I put on my uniform, no other options.

"C'mon GI," Teddy proclaimed as I came into the living room, uniformed, "We're burning daylight. Gotta' go."

"Where are we going?" I asked.

" You'll see, let's get after it."

Teddy drove to, of all places, The Pewter Pot. Teddy was not a Pewter Pot type guy. Strictly Lefty's. This was my surprise?

Margaret Mary shot up from her table when I walked through the door. My heart leapt, Skipped several beats, flipped, fluttered, as she crashed into me, all hugs and hellos, kisses and smiles. Much better surprise than the Pewter Pot. "How?" I stammered, "How did you get here? When did you get here?" These were not questions they were answered prayers.

"Teddy picked me up at the airport this morning. I couldn't get a flight yesterday. I was so mad. I wanted to be there when you got off the plane."

Margaret Mary held my arm tight and led me to a table in the back. Seated there, grinning like a cat, was Peter Rayburn, Father Peter Rayburn? Brother Peter Rayburn? Our Peter Rayburn. Teddy joined us, the gang was all here.

We sat and talked and laughed and caught up, hitting the highlights, saving the damage for later. Food arrived, food disappeared, coffee, tea and muffins came and went, hours passed. Then I saw Peter and Teddy exchange a look and Teddy said, "Dave, why don't you take my car and I'll catch a ride home with Peter?" He tossed me his keys. "See you laters" were spoken. Margaret Mary and I were alone at last.

The twinkle, the dazzle in her eyes winked off. She scooted her chair closer to mine and asked, "Are you alright? Really alright? Was it awful?"

"I'm okay, so far so good, you know? How about you? I need to know all about this Kendall guy. I need to know how he hurt you."

Her eyes filled with tears; her chin dipped to her chest. "Not now, alright? Not tonight. Later we will talk, I promise. Just not tonight."

"Where are you staying anyway?" I asked.

"At your house, silly. Your Mother is making up a bed down cellar by the pool table. She said I must stay with you but there will be no sleeping together under her roof without marriage."

"I did ask you to marry me once, remember?"

"You did," she recalled with her million-dollar smile, "and I loved you for it then and I love you for it now."

"That feels like a million years ago," I said.

"David, can we take a drive? Just around town, let's play the radio loud and go nowhere in particular for hours and hours."

So, we did, just like I hoped Danny H was doing.

The next morning when I woke up Margaret Mary was in the kitchen having coffee with my mother. Very domestic. I was in day three of my military uniform and needed to go buy some clothes. The topic was not addressed in the kitchen. Margaret Mary and I left to go shopping. I bought a winter warm and practical Navy Pea Coat, two pairs of jeans and two shirts at the Army-Navy store in Kearney Square. Next, we went to Pollards for a pair of loafers, underwear, socks and pajamas. Margaret Mary asked about my old clothes. She winced at the answer and changed the subject.

"David, can we go away somewhere? Just the two of us? I can stay two more days then I've got to go back for my mid-terms. That couch in your basement is lumpy and damp."

"Teddy said I could keep his car for as long as I need it. He's going back to Virginia tomorrow. How would you like to go up to York Beach?"

"Could we? That would be wonderful! Where would we stay? What will your Mother say?"

"My Mother will have to deal with it. We're not kid's anymore and I won't act that way. She said not under her roof, remember? We'll find our own roof in York Beach."

"And we can talk?"

"Yeah, and we can talk."

The next day we saw Teddy off at the bus station, made our excuses and said goodbyes to my parents and set off for York Beach. Margaret Mary sat very close to me on the ride up to Maine. Familiar sights rolled by. The closer we got to York Beach the more I longed to see it. Margaret Mary felt the same way. In an hour and a half, we were there. The day was clear and crisp and Autumn bright, the sky was an impossible blue and the glimpses we caught of the ocean even bluer.

When we first sighted Nubble Light off in the distance I felt a fond remembrance, a nostalgia alive and surrounding me, us, there, then and always. Even more than when I stepped into the living room on Glenmere Street two days ago I felt I was home, safe, and entire once again.

"Oh David, it's so beautiful, just like I remembered it," Margaret Mary acclaimed. I felt the same way but could find no words. Paradise, safe harbor from the storm, Sanctuary.

We checked into a musty, off-season, chilly room at the Sea Latch Motel, right across from Long Sands beach. The tiny room had a double bed, one night table with lamp, a tiny, rabbit-eared TV on a dresser and a phone booth sized bathroom. It was perfect.

Margaret Mary was staring at the double bed. In a low voice she said, "David, I can't, I'm not ready for, you know…"

I put my arms around her, looked her in the frightened eyes and answered, "Not an issue, not now. We have other matters to discuss, I have things I need to talk with you about. Let's take a walk on the beach, a long, long walk."

Mid-afternoon we stopped for lobster rolls and clam chowder, hot apple cider and cinnamon rolls. Then we walked those off as well.

"The first time he hit me I was so surprised I couldn't move or speak or do anything. Then he hit me again and I fell to pieces." We were back in our room now, long after sunset. The easy topics had passed. The painful ones were coming forward.

If Margaret Mary had sensed the extent of the inner rage and hatred I was feeling as she told me of Kendall and her time at Tuloc Meadows, I'm sure she would have stopped. She was, however, as deep into her shame and guilt as I was into my hatred and revenge.

She told me of the trial, how the other members of the group had lied, defended Kendall, slandered her, left her in ashes. She also told me of her therapy, long hours of understanding and compassion which were finally taking hold. She explained that she was in recovery, far from recovered, but working at it every day, none harder than she was right now. She told me how much Peter had helped, how much I was helping now, how much better she longed to feel and how much she needed me to understand.

I tried to say all the right things, make all the right promises, behave the right way, but inside I was cold as ice, hot as fire, hard as steel. I would find Kendall one day. I would not tell Margaret Mary when I did. I didn't care if it meant spending the rest of my life in prison. I would find Kendall one day.

Somewhere in the middle of her story I realized Margaret Mary was asleep in my arms. I knew it was late, after midnight and as stealthily as I could I reached over and clicked off the bedside lamp. We slept wrapped in each others' arms, fully dressed, atop the thin bedspread in our tiny, perfect room.

The next morning, we awoke, pancakes and eggs, coffee, toast, bacon and sausage hungry, starving. We drove around until we found an open diner and dug into all of the above. Sated, we planned our one more day in York Beach. Downtown York Beach was largely shuttered up and closed for the season. Fun-O-Rama, the bowling alley, the Goldenrod and Animal Forest Park were empty and silent, yet still perfect in their time and place. We strolled among the seasonally abandoned gift shops, snack bars, souvenir stands and restaurants. A small hardware store was open, we shopped, the proprietor was glad to see us. We returned to the beach where, as we walked, she asked about my time in Vietnam. I had to start somewhere, so I started with the men I served with.

"They were the best, Mags, are the best. Every day they risked their lives to save others, people we didn't even know, who didn't know us. They did it because it was the right thing to do."

"And you did too, right?" She asked.

"I did because they did. I was proud to be one of them, so lucky that I was assigned to that unit. Vietnam was a bad place, Mags. You could feel the evil, the fear and the hatred and the cruelty all around and it ate guys up. Both sides, good guys and bad guys did horrible things, acts of cruelty and despair, back and forth, day in and day out. Everywhere."

Saying these things out loud made my memories more vivid to me, my feelings more intense, my bewilderment more disorienting. Suddenly I had to stop, look around, struggle to remember where I was, who I was with, catch my breath and calm down. All the fear, anxiety, panic and disbelief I could not allow myself to feel in Vietnam washed over me. Mags saw me swoon and I sank to the sand; I couldn't stop my tears. She held me. I kept crying.

Later that night we lay once again on our bed, wrapped in each other's arms quiet, calm, safe for this time. Tomorrow we would go back to Lowell. Margaret Mary would go back to Chicago. And for the first time since I got off the plane three days ago in Boston, I felt my military career was over. I was out of the Army. I was home.

Wars are not paid for in wartime, the bill comes later.
Benjamin Franklin
EVERYDAY POWER

Epilogue

There is one more tale to tell, one more story left unfinished before I close the chapter of the life I lived in Vietnam. It is not a happy tale, nor is it complete, but here it is, as best I can tell it

The Unhappy Ending of Harry Giles

Harry kept the promise he made to his wife when we returned to Phu Bai after our thirty-day leave. He went to Major Durant and asked to be taken off flight status. Major Durant understood fully, and Harry was transferred to the 22nd Surgical Hospital as a ward medic. He worked with the most severely wounded, gunshot wounds and multiple frags and initially did a great job. Then Harry started being Harry. We saw less and less of him in our unit area. He was hanging around with a new crowd, naturally enough, medics up at the hospital. When we did see Harry, he seemed worn down, subdued, not the about to be out of control puppy we had come to know.

One night, returning from chow, I ran into Harry.

"Harry, how's it going?" I was glad to see him. He didn't seem to recognize me at first.

"Hey, Dave," he answered, the first time he ever used my name without "Yo" in front of it. He didn't look me in the eye either.

"You okay, Harry?"

"I'm good, man, goin' to work."

"You look like you should be going to bed."

"Tired, man, gottta' go. See you around."

And he shuffled off toward the wards. Most everybody I knew was tired most of the time when we weren't scared, drunk or faking it. I didn't give Harry another thought until a few days later when Don said, "Hey did you guys hear about the big raid up at the hospital last night?"

I hadn't, we hadn't, Don continued, "MPs rounded up a bunch of guys. Took them away. Nobody's saying much but it seems to be about a lot of medical supplies going missing."

I immediately thought of Harry.

"What about Giles?"

"Don't know," Don answered, "it's all pretty hush-hush right now."

So I went to look for Harry. His bunk was empty. His gear was still there. I went to the ward where he worked. Not on duty was all they would say. Not around the compound either. My search was cut short by a second up mission. We were flying to Danang to pick up supplies. I knew who to ask when we got there.

"Ferrara-ara-ay, get in here," Command Sergeant Major Hoadwonick barked. "Did you come here to annoy me or to ruin my otherwise perfectly delightful day?"

"No Sergeant Major, I came here to ask you a question."

"Is it a one-drink or a two-drink question?"

"Probably two Sergeant Major, but it is a little early in the day."

"Not in Honolulu it isn't, Fernarier. What is your question?"

"I came to ask if you know what happened to Harry Giles, Sergeant Major, one of our medics."

"I know who Giles is, Furry-er-er. Close that door."

I did. He continued, in a lower tone.

"Giles is under arrest along with four other shit birds from the Surg."

"May I ask why, Sergeant Major?"

"You just did, and no, you may not. I will tell you that Giles has run out of tolerance, and he will not be coming back to the Surgical hospital."

"Please Sergeant Major, Harry is my friend. I'd really like to know."

Sergeant Major Hoadwonick considered me for an overly long minute. He clasped his hands on his desk and explained, "Someone has been stealing more than considerable amounts of morphine from the hospital's supplies. We didn't notice at first because of the high number of casualties coming through. Two days ago, one of the shit birds now in custody overdosed. Morphine. Morphine intended for the wounded. When he came to, he gave up his buddies in a heartbeat. Giles was one of those buddies."

I started to say something, Sergeant Major Hoadwonick held up his hand, "We tested the shirt birds for drugs. Harry was positive. He confessed."

"What's going to happen to him?"

"Something very unpleasant and very far from here. Stealing medication from the wounded is the bottom of the barrel. This bunch is going to become very familiar with the bottom of the barrel."

Then his demeanor changed, the closest I ever saw it verge upon sympathy, "I have the greatest respect in the world for the medics. What they go through is above and beyond every single day. There was only one other medic involved beside Giles. I will put in a very small, good word for both of them when they hammer the others. Your friend is going to prison, Long Binh first, then Leavenworth."

I was speechless. Sergeant Hoadwonick continued, "What I am telling you here is not to go beyond this office. There is an ongoing investigation and you do not want to rat fuck my ongoing investigation do you, Specialist Ferrier?"

I didn't. And neither I, nor anyone in the 571st Dustoff ever saw Harry Giles again.

If you enjoyed this book, I would really appreciate a brief review. Your help in spreading the word is greatly appreciated and reviews make it easier for readers to find books.

THE MOUNTAINTOP SERIES:

Follow the adventures of Dave and his friends in the first two books in the series:

Available from Amazon:

BORN ON A MOUNTAINTOP (Book One of The Mountaintop Series):
https://www.amazon.com/dp/B09ZYW5FXT

RAISED ON ROCK (Book Two of The Mountaintop Series):
https://www.amazon.com/dp/B0CC3MHMY2
&
Battle Press
https://battlepress.media/

Book Four, Wired For Sound, and Book 5, California Dreaming coming in 2024.